Mark Levi

WHY CATS LAND ON THEIR FEET

And 76 Other Physical Paradoxes and Puzzles

PRINCETON UNIVERSITY PRESS

PRINCETON AND OXFORD

Library of Congress Cataloging-in-Publication Data
Levi, Mark, 1951–
Why cats land on their feet, and 76 other physical paradoxes
and puzzles / Mark Levi.
  p. cm.
ISBN 978-0-691-14854-0 (pbk. : alk. paper)
1. Science–Miscellanea.   I. Title.
Q173.L545 2012
530–dc23   2011045728

British Library Cataloging-in-Publication Data is available

This book has been composed in Times

Printed on acid-free paper ∞

Typeset by S R Nova Pvt Ltd, Bangalore, India
Printed in the United States of America

1   3   5   7   9   10   8   6   4   2

TO

Olga,

Kyra,

Eric,

Max,

Nika,

Vicki,

Jose,

and

Ryan

CONTENTS

ACKNOWLEDGMENTS

I AM GRATEFUL to Paul Nahin for several very useful comments, in particular for pointing out the references to the Braess paradox, to Carole Schwager for numerous improvements of the exposition, and to Vickie Kearn, whose encouragement kept me going. I gratefully acknowledge support by the National Science Foundation under Grant No. 1009130.

WHY CATS LAND ON THEIR FEET

1

FUN WITH PHYSICAL PARADOXES, PUZZLES, AND PROBLEMS

1.1 Introduction

A good physical paradox is (1) a surprise, (2) a puzzle, and (3) a lesson, rolled into one fun package. A paradox often involves a very convincing argument leading to a wrong conclusion that seems right, or to a right conclusion that seems wrong or surprising. The challenge to find the mistake—or explain the surprise—may be hard to resist. A joke heard back in the Cold War years claimed that the West could impede Soviet military R&D efforts by scattering leaflets containing puzzlers and brainteasers over the secret Siberian weapons research facilities. Times have changed, and these same brainteasers now are used in hiring interviews. As a Soviet propagandist would have said: either way, they are a capitalist tool.

Resolving a paradox is not only fun; it also trains intuition, logic, and critical thinking. One becomes a better lie detector by resolving paradoxes. A good paradox also teaches caution and humility by showing us how easy it is to go wrong even in relatively simple matters of elementary physics. It is liberating to know that some very smart people have made mistakes in seemingly simple matters of basic physics. Other fields—such as astronomy, biology, medicine, economics, climate, politics, and media—deal with more complicated objects than physics,[1] offering even more room for mistakes there. In addition, some 'mistakes' can be beneficial, at least temporarily.

My main reason in writing this book is to share the fun of imagining how things work. These paradoxes also teach the gist of some physics without the pain of mathematics.[2]

The puzzles in this book deal with physics—a subject that walks on two legs, one being mathematics, and the other, physical intuition. Unfortunately, in school the subject is often presented with a severe limp.

A musical analogy. If music were taught the way physics often is taught, we would learn the notes but not the melodies they produce. For too many students of physics, the subject is reduced to a collection of formulas that must be matched to a problem at hand. Not surprisingly, many intelligent students are turned off.

Intuition should come first. Exercise of physical intuition is one *practical* benefit of this book's puzzles. All too many physics courses give short shrift to intuition, emphasizing instead a search for the formula that fits the situation. Examples in this book go in the opposite direction: I tried for a minimum of formulas and a maximum of intuition. The discussion of the spinning top is an example, where I give a formula-free explanation of why the spinning top stays

---

[1] This is not a statement on the relative difficulty of various sciences. I am simply referring to the fact that a physicist deals with much simpler objects (e.g., crystals) than a biologist (e.g., a cell).

[2] I refer to "pain" with tongue in cheek—mathematics is of course indispensable and beautiful to me, at least, since it's my job.

upright. It takes quite a few years of study in mathematics and physics to learn to write differential equations for the motion of a spinning top and to see how to deduce stability from these equations. And at the end of this long study few students end up with an intuitive understanding of why a spinning top stays up. The most powerful tool—our physical intuition—ends up unused.

1.2 Background

Much (but not all) of this book should be accessible to readers without formal background in physics. All physical concepts used are explained in the appendix. Mathematics in this book does not go beyond algebra, with a couple of exceptions where calculus is used. Even there, the reader who is willing to take a little math on trust should not be snagged by these references.[3]

Attraction to anything surprising is a basic instinct in most living creatures, or, at least, most mammals. By driving us to explore, the instinct helps us survive—with some exceptions, such as Darwin Prize winners or the heroes of Jackass. The same instinct that drove Einstein to his great discoveries also drives a curious child to see what's inside a mechanical clock. It even drives puppies and cubs to explore. In some people this instinct is so strong it can survive the educational system.

1.3 Sources

This book grew out of a collection of puzzles I started long ago on my father's advice, after I showed him one that

---

[3] I am referring here to the sling problem on page 93 where the rock reaches infinite speed after one second.

occurred to me after a high school class on the capillary effect (page 128). Although I invented some of this book's puzzles,[4] *it is most likely that others thought of them or of something equivalent before I was born.* When I know the author or the origin of a puzzle, I make a reference.

Literature. Fortunately, much of the essence of basic physics can be understood, and enjoyed, without (m)any formulas, as some excellent popular books demonstrate. Among these are Walker's *The Flying Circus of Physics*, Epstein's *Thinking Physics*, Jargodzki and Potter's *Mad about Physics*, and Perelman's classic *Physics for Entertainment*. Unfortunately, Makovetsky's delightful book *Smotri v koren'* (a loose translation: "Seek the essence"), which sold over a million copies in the former Soviet Union, does not seem to have been translated into English. Minnaert's *The Nature of Light and Color in the Open Air*, dedicated to optical phenomena in nature, will never age and will give pleasure to any curious individual lucky enough to open it.

---

[4] For example, 2.1, 2.3, 2.4, 3.1, 3.2, 3.5, 3.6, 4.1, 4.2, 4.4–4.6, 5.3–5.8, 6.6, 6.7, 6.10–6.12, 8.2, 8.5, 8.6, 9.4, 11.1, 12.3, 13.2, 14.6, 14.8.

⊠ 2 ⊠

OUTER SPACE PARADOXES

2.1 A Helium Balloon in a Space Shuttle

Problem. Two astronauts, Al and Bob, are strapped to the opposite ends in a space capsule, as in Figure 2.1. Al is holding a large helium-filled balloon, and everything is at rest. Now Al pushes the balloon, which begins to drift toward Bob. Which way will the capsule drift, as seen by an observer hovering in space outside the capsule? Since the astronauts are strapped to the walls, let's consider them part of the capsule.

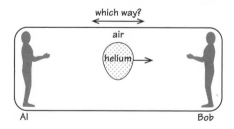

Figure 2.1. Which way does the capsule move after Al pushes the balloon?

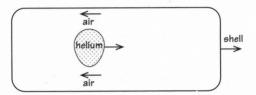

Figure 2.2. Motion as viewed from the capsule's (and Al's) reference frame.

A reasonable guess. When Al pushes right, the balloon pushes him back, according to Newton's "action equals reaction" third law. And since the balloon pushes Al left, he, and the shuttle, will drift left. Is this correct?

Answer. Actually, no: the capsule will drift right as well!

An explanation via center of mass. The center of mass of the entire system (the capsule and the contents) is fixed, because there are no external forces (all the concepts in this sentence are explained in the appendix, (pages 169–73). Now the motion inside the capsule, *from Al's point of view*, is sketched in Figure 2.2. The balloon has a lot less mass than the air it displaces, and so, from Al's point of view, the center of mass moves *left*. But the center of mass of the whole system is fixed in space, since there are no external forces. Therefore, Al himself, and the capsule, move right from the external observer's point of view.

Our mistake has been in paying too much attention to the balloon and not enough attention to the more massive air that moves left to replace the moving balloon.

An equivalent explanation via linear momentum. As explained in the appendix (pages 169–73), the fact that the center of mass stays put is equivalent to saying that the linear momentum remains zero. Now from Al's viewpoint,

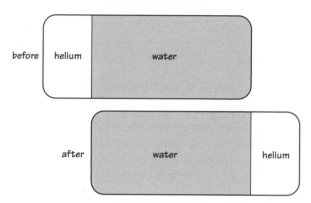

Figure 2.3. Water stays in place; the nearly massless shell moves right.

the displaced air moves left. This signals that Al himself (and the shuttle) are moving right, to cancel the leftward motion of the air and to keep the linear momentum at zero.

The intuitive feel of all this becomes particularly clear by taking the mass ratio to the extreme, as in Figure 2.3, where helium/air is replaced by helium/water. Since the water carries virtually all the mass, it stays essentially in place. And thus, as the helium is transferred to the right, the shell (whose mass we neglect) will move in the same direction to accommodate the helium bubble.

A nagging doubt. The above answer is correct. But didn't I prove the opposite on page 5, in the "reasonable guess"? Where is the mistake in that "proof"?

Answer. The mistake was in not considering *all* the forces on Al. The force he feels from the shell was forgotten! In fact, his push on the balloon is transferred via the air to the shell, which pushes Al in the back. Surprisingly, this push is stronger than his push on the balloon. In effect, by kicking the balloon, he kicked himself in the rear even harder! This is quite a surprise (especially for Al). How exactly does this happen? The next paragraph gives an intuitive feel of this.

How to kick oneself in the butt (with one's own knee)? This short paragraph explains why the shell pushes *Al* in the back harder than he pushes the balloon. To make things easier to "feel," let us temporarily change the problem: imagine that instead of helium, the balloon is filled with air. Now as Al pushes the balloon he simply reshuffles the air inside the capsule.[1] By rearranging the air inside the shuttle he is not changing the air's center of mass, and thus the shell and Al will not move. By Newton's first law, the net force on *Al* is zero: *his palms and his back feel equal and opposite forces*. Now what changes if the balloon is filled with helium? The balloon now has less inertia and thus is easier to accelerate. So Al's palms feel lesser force than before. This explains why the force on his back is greater if the balloon is filled with helium.

A childhood memory. This paradox reminds me of a mistake I had made as a child, obsessed by an overwhelming desire to fly. One day inspiration struck. Hopping into a chair, I grabbed the seat and started pulling it up with all my strength. The brilliant idea was to make the seat push me upward, thus turning me into a human rocket (with the chair legs flaring out like Sputnik's antennae). But the takeoff failed. I had overlooked the fact that my arms were (indirectly) connected to my rear end (something that now seems both obvious and fortunate), thus pushing it down into the seat and canceling the lift of my hands. I was pushing on the chair with *two* ends of my body and overlooked one of

---

[1] We take the balloon's skin to be massless.

these. The similarity with the capsule paradox should now be obvious. Just as in the first discussion, I did not account for all the forces. The balloon pusher is connected to the balloon not by his hands alone, but by his back as well, via the shell and the air; the force of this other connection was overlooked in the initial discussion.

2.2 Space Navigation without Jets

Problem. Can a satellite change its orbit around Earth without using jets, solar wind, or other means of propulsion? Using solar panels to gather energy from the Sun is allowed.

A hint. Try using the fact that the gravitational force depends on the distance from Earth. And the satellite must not be just a point mass.

Answer. The simplest satellite that will work consists of two masses connected by a cable. A motor, driven by a solar battery, can change the length of the cable. The satellite starts out orbiting Earth and tumbling, as illustrated in Figure 2.4.

It turns out that we can raise or lower the orbit of this satellite by appropriately varying the length of the cable! To see how, let's assume that the satellite starts out orbiting and

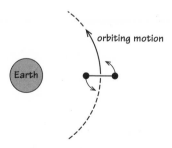

Figure 2.4. The "dumbbell" satellite of adjustable length.

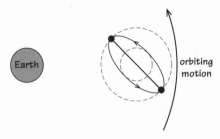

Figure 2.5. To descend to a lower orbit, shorten the cable when it points toward Earth.

Figure 2.6. Gravitational pull is stronger on A than on B, putting tension on the cable.

tumbling as shown in the Figure 2.4. Our task is to (say) descend to a lower orbit. To do so we follow the instructions in Figure 2.5: shorten the cable when it points at Earth, and lengthen it when it is near the perpendicular direction. By doing this repeatedly over many tumbles we will cause the satellite to descend. To ascend, we do the opposite.

An explanation in a nutshell. Due to the tumbling motion, the cable will be under centrifugal tension. But not only centrifugal: this tension varies slightly, as Figure 2.6 explains, due to a tidal-like effect.

As Figure 2.6 explains, the tension is greatest when the cable points directly at Earth since the masses A and B are then at different distances from Earth, with the resulting difference in gravitational pulls upon A and B stretching the cable.[2] Now this variation in tension can be used to speed up the tumbling: by shortening the cable when the tension is greater and by letting it out when the tension is smaller we do work. This work goes into faster tumbling.[3] And the increase in angular momentum of tumbling means a decrease of orbital angular momentum,[4] since the total angular momentum is conserved. And lesser orbital angular momentum means that the satellite is in the *lower* orbit, as I show after two paragraphs.

An explanation by smearing. An easy alternative way to understand why the tumbling angular momentum of the satellite increases in Figure 2.5 is the following. Imagine that the masses of the two balls of the dumbbell are smeared along their paths shown in Figure 2.5. So we replaced the moving satellites by a "wire hoop." The point is that this hoop is tilted relative to the direction toward Earth, and thus it feels torque due to the gravity being stronger closer to Earth. This torque tries to turn the hoop counterclockwise, that is, it *increases* the hoop's angular momentum—in agreement with our initial reasoning. (A similar idea of "smearing" is applied to explain a gymnastics skill on page 60).

Here are a few more details skimmed over in the above explanation, including the key point that the higher orbits have more angular momentum. The total angular momentum

---

[2] The same effect is responsible for the tidal stretching of Earth along the Moon–Earth line.

[3] Swings are pumped up the same way: we lift a part of the body when the g-force is greater and lower it when the g-force is less, thus doing work which goes into the swinging motion (discussed in more detail on pages 57–58). This recipe ("buy high, sell low") works for increasing kinetic energy, but would spell ruin if investing.

[4] See page 175 for the background on the angular momentum.

of the satellite with respect to Earth's center remains constant whatever we do with the cable:

$$M = M_{\text{tumbling}} + M_{\text{orbital}} = \text{const},$$

since the gravity points straight at the center of Earth, thus exerting zero torque upon the satellite.[5] So by increasing M_{tumbling} we decrease M_{orbital}, the angular momentum around Earth. But this momentum is directly related to the radius r of the orbit via

$$M_{\text{orbital}} = k\sqrt{r}, \qquad (2.1)$$

where $k = m\sqrt{GM}$; here m is the mass of the satellite, G is the universal gravitational constant, and m is the mass of Earth.[6] According to the formula, decreasing M_{orbital} causes a decrease in r. This confirms the above claim that by speeding up the tumbling we decrease the orbital angular momentum; the satellite descends.

Other maneuvers. One can actually do more—namely, one can also change the eccentricity of the orbit; how to do it is left as a challenge to the reader.

---

[5] This is not exactly true because Earth is not a perfect sphere; let's ignore this complication.

[6] Indeed, for the circular orbiting motion of radius r, the definition says:

$$M_{\text{orbital}} = mvr; \qquad (2.2)$$

v is the speed, determined by r via Newton's second law. According to this law, the centripetal acceleration (v^2/r) is determined by gravity via

$$\frac{mv^2}{r} = \frac{GmM}{r^2},$$

which gives $v = m\sqrt{GM}/\sqrt{r}$. Substituting this into (2.2) gives (2.1).

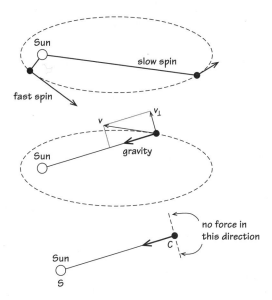

Figure 2.7. (a) No force is acting perpendicular to the line SC. Does $v_\perp$ stay constant? (b) The Sun–comet line turns faster when the comet is closer to the sun.

2.3 A Paradox with a Comet

In discussing motion of projectiles, a standard argument goes that the horizontal speed of the projectile remains constant since there are no forces acting in the horizontal direction (air resistance is ignored). Here is an attempt to extend this reasoning to outer space.

The Sun pulls an orbiting comet but does not try to "spin" the comet around the Sun: the gravitational force points straight into the Sun at all times (see Figure 2.7). Put differently, the force of solar attraction has zero component in the direction perpendicular to the Sun–comet line: $\mathbf{F}_\perp = 0$. By Newton's first law (zero force means no change in

13

speed) I conclude that, since the force in the perpendicular direction is zero, the velocity $v_\perp$ in that direction stays constant. And then come second thoughts. The comet's angular momentum[7] is conserved:

$$M = mv_\perp \cdot r = \text{const},$$

where r is the distance from the comet to the Sun's center. And since r changes as the comet moves in its elliptical orbit, $v_\perp$ must adjust to keep the product $v_\perp \cdot r$ constant. Which (if either) of these two thoughts is correct?

Solution. The mistake is in the first argument: I misstated Newton's first law. The law is valid only in an inertial frame.[8] Implicitly, I treated SC as the coordinate axis of an inertial frame. But this frame is certainly not an inertial one, since the axis rotates.

2.4 Speeding Up Causes a Slowdown

Question. A spaceship is in a circular orbit around a planet. Wishing to go to a higher orbit, the astronaut activates the jets, propelling the ship forward. Once the ship is in a higher circular orbit, the jets are shut off. The ship was pushed approximately in the forward direction during the burn; does it now move faster than it did initially?

Answer. The ship actually slowed down.

An explanation. This fact may seem less strange if we recall that while biking uphill, you may be pedaling hard and yet slowing down. This is what happens to the spaceship:

---

[7] See page 175 for the definition and the background.

[8] That is, in a frame which moves with no acceleration and with no rotation. For more details on Newton's law see the appendix, page 161.

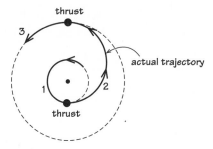

Figure 2.8. Forward propulsion causes a slowdown. In this illustration two short forward bursts are applied as shown, resulting in a higher orbit, with a slower motion.

going to a higher orbit is like biking uphill. The propulsion was spent not on speeding up, but rather on overcoming the gravitational pull. The ship gains potential energy and it loses kinetic energy. But it gains more than it loses.

Precisely how does the orbital speed v depend on the orbit's radius r? The answer turns out to be

$$v = \frac{k}{\sqrt{r}},$$

where $k = \sqrt{GM}$; here G is the universal gravitational constant and M is the mass of the planet. When in a circular orbit, Newton's second law $ma = F$ states that the the centripetal acceleration[9] $a = v^2/r$ is caused by the gravitational force $F = GmM/r^2$:

$$m\frac{v^2}{r} = \frac{GmM}{r^2},$$

---

[9] That is, the acceleration toward the center. See page 179 for details and background.

15

where m is the mass of the satellite and G is the universal gravitational constant. We solve for the orbital speed v:

$$v = \sqrt{\frac{GM}{r}}.$$

This formula indeed shows that by climbing, that is, by increasing r, the satellite slows down.

❋ 3 ❋

PARADOXES WITH SPINNING WATER

Archimedes discovered his famous law: the buoyancy force exerted by water upon a submerged body equals the weight of the water displaced by the body.[1]

In a rotating world, such as Earth, Archmedes' buoyancy law acquires a twist (no pun intended), with some surprising manifestations. One such surprise is the paradox of the floating cork described next. Another is the iceberg paradox (p. 25).

3.1 A Puzzle with a Floating Cork

The experiment. An amusement park with a spinning swimming pool would be a dream of any child, even one

---

[1] The following thought experiment explains why this law is true. I want to explain why a bowling ball lying on the bottom of the pool feels the buoyancy force equal to the weight of water that this ball displaces. As a thought experiment, imagine replacing the ball with the same shaped water ball. This water ball will hang still, since the water is assumed to be still in the pool. We conclude that the gravity exactly cancels the buoyancy for the water ball. So for the water ball at least, the buoyancy equals the weight of water. But the buoyancy force depends only on the shape, and thus is the same for the bowling ball. This proves Archimedes' law. In short, Archimedes' law boils down to two things: (1) still water stays still if undisturbed, and (2) the buoyancy force upon a body depends only on the body's shape, but not the material it's made of.

trapped in the body of an adult. With the image of such a pool in the back of my mind, I was planning to demonstrate to my calculus class that the water surface in a spinning bowl takes the shape of a paraboloid. A large salad bowl filled with water and placed on the spinning disk of a record player worked very well. In a minute or so the water settled into the 33 rpm rotation, revolving with the bowl as a single solid. The water surface became a shiny, perfect paraboloid.

I then placed a cork on the sloped surface, just out of curiosity. I had expected the cork to stay on the incline—it would have been a fascinating thing to watch, and perhaps to imagine myself floating in a spinning pool—what an amusement ride it would have been! The cork, however, did the unexpected: it slowly drifted toward the bottom of the paraboloid until it came to rest there. Maybe it's the air resistance, I thought. Still unsure, I covered the bowl with clear plastic wrap. With the cork floating near the wall, I turned the motor back on. The same thing happened again! So the air wasn't the cause: the air settles into rotation sooner than the water. The internal motions die out much faster in the air than they do in the water. The air is actually more viscous than water if one measures viscosity relative to density.[2]

Question. Why did the cork drift down?

The explanation of the drift. Let us put ourselves in the rotating frame of the turntable. Figure 3.1 shows the blob B of water similar to the one displaced by the cork. In our rotating frame, the blob is at rest. This means that the

---

[2] Such relative viscosity is called the *kinematic viscosity*, defined as the ratio of viscosity to the density—in the standard notation, $\nu = \mu/\rho$, pronounced "new." Back in grad school, learning fluid dynamics from Joe Keller, we were told that if someone greets us with the words "what's new?" we can now reply "mu over rho."

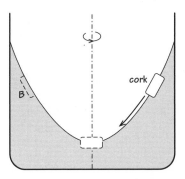

Figure 3.1. The buoyancy forces on the cork (right) and on the dotted blob of water (left) are the same. But the centrifugal force on the cork is weaker since the cork's center is closer to the axis—hence the drift.

centrifugal force[3] balances out the horizontal component of the buoyancy. Now imagine the blob B slowly swelling and becoming a cork, while keeping its underwater shape the same. In the process, the blob's particles will be getting closer to the axis. The centrifugal force on them will thus weaken, while the buoyancy will not change. This imbalance pushes the cork toward the axis.

3.2 Parabolic Mirrors and Two Kitchen Puzzles

Parabolic mirrors. Telescopes' mirrors are shaped as paraboloids (the paraboloid is the surface swept by a parabola spinning around its symmetry axis, as in Figure 3.1). The paraboloids are so useful because they collect a beam of rays parallel to the axis into a single point.[4]

---

[3] See the appendix, page 180, for the discussion of the centrifugal force. The centrifugal force is a fictitious force which arises because the reference system is not inertial.

[4] The microphone antennas for eavesdropping, satellite dishes, and other antennas are also shaped as paraboloids for the same reason. The sensor is then placed at the focus of the paraboloid, thus collecting all the "rays" which bounce off the antenna.

Only paraboloids among all surfaces have this focusing property. Now by a wonderful coincidence, nature provides a simple way to create paraboloidal shapes: the surface of spinning liquid, as in Figure 3.1, automatically becomes paraboloidal.

By allowing molten glass in a spinning container to cool slowly, we get a nice paraboloid, without any machining. Nature is an analog computer and a lathe rolled into one when it comes to paraboloids.

An edible puzzle. Set the liquid gelatin mix in a bowl on a spinning turntable, and let it spin until it sets. The surface of coagulated gelatin will be a nice paraboloid. Your friends may be puzzled (and perhaps a bit worried), possibly thinking that you must have spent hours meticulously scooping out the indentation to such incredible smoothness. Later you can fill the paraboloid with whipped cream and feed it to your guests.

A culinary application of Taylor-Proudman's theorem. Here is a way to create interesting color patterns in gelatin. Set a glass with liquid gelatin on the spinning turntable, giving it a few seconds to set rotating with the turntable, and then pour a tablespoon or two of differently colored gelatin. The two fluids will mix in a surprising way: the poured-in gelatin will form a curtain-shaped roll. Wait until the remarkable pattern coagulates. Showing this shape to your friends may puzzle them (but hopefully not decrease their number), although by now they will probably guess that rotation played a role.

This strange mixing is due to the gyroscopic effect in fluids. The phenomenon is captured by the Taylor-Proudman theorem, which states, roughly speaking, that a rapidly rotating fluid acquires a directional "stiffness," behaving as if it were made of "toothpicks" parallel to the axis of rotation. The faster the rotation (compared to the internal speed of the fluid), the greater this stiffness.

This effect plays a role in the motion of the atmosphere and of the oceans. A more detailed discussion (the intuitive part of which requires no calculus) can be found in the classical book by G. K. Batchelor, *An Introduction to Fluid Dynamics*.

3.3 A Cold Parabolic Dish

Problem. Imagine water in a container standing on a steadily spinning platform, Figure 3.2. The surface of spinning water turns out, no pun intended, to be a paraboloid, as the figure illustrates. That is, we get a parabola $y = kx^2$ by slicing the surface by a vertical plane through the spin axis. The steepness of the parabola is determined by k, which depends on the angular velocity[5] ω of the turntable and on

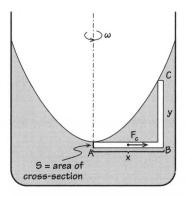

Figure 3.2. Toward an explanation of the parabolic shape of the surface.

---

[5] See appendix, page 178.

the gravitational acceleration g:

$$k = \frac{\omega^2}{2g}. \tag{3.1}$$

On the Moon, where $g_{\text{moon}} \approx g/6$, this paraboloid will be stretched six times vertically compared to one on Earth; on Jupiter the paraboloid would be about 2.5 times flatter. And if you reset your turntable speed from 33 to 78 rpm—a little more than doubling—then k will increase almost sixfold. The same effect as going to the Moon, but cheaper to achieve.

Why does water choose to have the shape of the paraboloid out of the infinity of all shapes? Here is a quick calculus-free explanation. The only background needed is the fact that the centrifugal force upon a point mass m spinning in a circle or radius r is

$$m\omega^2 r, \tag{3.2}$$

as explained in the appendix on page 180–81.

The problem is to find the depth at any distance x from the axis (Figure 3.2). Here is a simple key to the solution.

The centrifugal effect upon the column AB creates an elevated pressure at B; this pressure holds up the column BC:

$$p_{AB} = p_{BC}. \tag{3.3}$$

On the one hand, the pressure p_{AB} at B is given by the centrifugal force F_c of the horizontal tube AB divided by the cross-sectional area S of the tube: $p_{AB} = F_c/S$. Now the centrifugal force $F_c = m\omega^2 r$ (as explained on page 181); here the mass of the tube is $m = $ density $\cdot$ volume $= \rho x S$, and $r = x/2$ is the distance of the tube's center of mass to the axis of rotation. Thus

$$p_{AB} = \frac{F_{AB}}{S} = \frac{\rho x S \omega^2 x / 2}{S} = \frac{1}{2}\rho\omega^2 x^2.$$

On the other hand, the pressure of the vertical column BC is $\rho g y$, where y is the depth:

$$p_{BC} = \rho g y.$$

Substituting the expressions for the pressure into (3.3), we get

$$\frac{1}{2}\rho\omega^2 x^2 = \rho g y$$

or

$$y = \frac{\omega^2}{2g}x^2,$$

as claimed.

By the way, the density ρ canceled out in the final result; only ω and g matter for the shape. Mercury or water surfaces will have the same shape, other things being equal. If you want to build a telescope mirror by spinning a molten material, density is one fewer thing to worry about.

3.4 Boating on a Slope

Question. Imagine again the water spinning in a bowl, as in Figure 3.3. A toy boat with remote control is floating on the surface. The operator wants to make the boat stay off center and at a fixed place relative to the ground, as shown in the figure. Which way should he point the nose of the boat? And which way should he steer: straight, right, or left?

Answer. The bow must be pointed in the direction 2, since the boat needs to have velocity opposite that of the flow, and

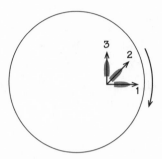

Figure 3.3. Which way should the boat point in a spinning container so that it stays in a fixed position relative to the ground?

it also needs some thrust away from the center so as not to slide down the slope: there is no centrifugal force to keep the boat on the slope since the boat is at rest.

3.5 Navigating with No Engine or Sails

Question. You find yourself in a boat in a rotating pool. Can you navigate without oars, propellers, or sails? Air resistance is to be ignored.

Answer. To move toward the center, *stand up*. Now, "up" in the rotating frame is not the same as the "up" on the ground: standing up you will be tilting toward the axis, and thus getting closer to the axis. The centrifugal force thus decreases, causing drift toward the axis.[6] To move away from the axis, lie down on the bottom of the boat, thus maximizing your distance to the axis and thereby increasing the outward drift.

---

[6] Your head will be closer to the axis of rotation than the rest of the body, and you will literally feel light-headed and heavy-footed.

Question. My buoyancy is barely positive. Will I float in a rotating pool?

Answer. Not necessarily. The legs have higher density than the chest, due to the air in the lungs, so I tend to float in a vertical position (with my head up, hopefully). Imagine now that I float near the edge of the rotating pool where the water surface is sloped steeply; my body, perpendicular to the surface, will then be almost horizontal. This puts my legs farther from the axis and makes the centrifugal force on them greater; this may win over the buoyancy, dragging me under. But I could try saving myself in several ways. First, I should try to float with my feet on the surface, pointing "downhill." Alternatively, I can sink to the wall, crawl to the bottom, and then crawl along the bottom to the center (all the while trying to move with my feet forward), whence the centrifugal force will be reduced to naught and I will ascend.

3.6 The Icebergs

Question. Do icebergs feel Earth's rotation? (Of course, the icebergs are affected by currents and winds which themselves are affected by Earth's rotation, but I am not talking about those indirect effects.)

Answer. Earth's rotation results in a force pulling the icebergs toward the equator. The "drifting cork" discussion (page 18) explains why. Here is the explanation, adapted for the iceberg. Imagine that, instead of being brought to its current location from elsewhere, the iceberg was created by freezing a chunk of water it displaces. The water swells, becoming the iceberg, with a part of it rising above the surface. But this increases the average distance of this water to Earth's axis. And the centrifugal force increases farther

25

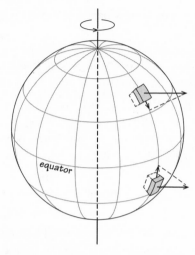

Figure 3.4. Icebergs are pulled toward the equator due to Earth's rotation.

from the axis. We conclude that the iceberg feels a little extra centrifugal force in the direction away from the axis! This force tries to drag the iceberg toward the equator, as Figure 3.4 explains. How significant is this force in practice? A rough estimate that I do not want to bore (and perhaps annoy) you with shows that for an iceberg of horizontal extent ≈10 km and of thickness 0.2 km the speed of steady drift caused by this force would be on the order of magnitude of 1 m/sec: at this speed the drag would balance the centrifugal force. One meter per second is a walking speed, and not that slow even compared to the speed of some ocean currents. There is a catch, however: you would need a year or so to achieve this speed starting from rest—the acceleration is that small, as it turns out. On the other hand, some icebergs last for a year or even two, giving enough time for this effect to show itself. Indeed, if we take the average speed as

0.5 m/sec, then over 1 year $\approx 3.15 \cdot 10^7$ sec $\approx \pi \cdot 10^7$ sec[7] we get the distance of about $(\pi/2)10^7$ m, or about 15,000 km— about the distance from a pole to the equator! I found this somewhat shocking. Of course, I was ignoring the much stronger effects of the wind and currents. But this centrifugal effect is weak yet relentless; the winds/currents are strong but variable. It is not clear (not to me, at least) whether the weak force biases the icebergs toward the equator.[8]

---

[7] I learned from Tadashi Tokieda the nice fact that 1 year is $\approx \pi \cdot 10^7$ sec.

[8] For more insight into this question more information about the currents and the winds is needed. One can make up theoretical examples of currents in which the weak drift has no effect, and others with which the drift has a huge effect. With more information available about the currents/winds, the question becomes one of dynamical systems and can be resolved under some idealized assumptions. A computer experiment may help a lot, especially for the currents unmanageable by theoretical tools. Real currents probably fall into that class.

※ 4 ※

FLOATING AND DIVING PARADOXES

4.1 A Bathtub on Wheels

Here is a twist on the "space ball" problem, this time on Earth.

Question. A toy boat bobs at one end of a tub with water. The tub is mounted on perfectly frictionless wheels and stands on a perfectly smooth floor. All is at rest. Using the remote control, you make the boat travel from one end of the tub to the other. Gradually all motion stops. Which way did the tub move from its initial position?

The masses of the boat and of the tub are m and M, respectively; the length of the tub is L.

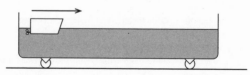

Figure 4.1. How far will the tub move after the boat travels from left to right?

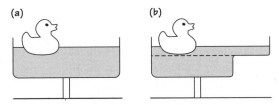

Figure 4.2. Will the dishes tip over when the ducks are slowly removed?

Answer.

$$\text{distance} = \frac{m}{m + M} L$$

is wrong. Actually, the tub will end up exactly where it started.

An explanation. According to Archimedes, the boat's mass is the same as the mass of displaced water. So moving the boat from one place to another amounts to swapping two equal masses, as Figure 4.1 illustrates. But swapping two equal masses does not change the center of mass of the system water + boat relative to the tub (C.M. hereafter).[1] But C.M. does not move relative to the ground, either, since no external forces are applied to the tub.[2] Hence the tub will not move.

Question. A round dish filled with water is balanced on a narrow support, Figure 4.2a. A rubber duck is floating near the edge of the dish. You slowly remove the duck. Which way will the dish tip—or will it?

Answer. The dish will not tip, but rather will remain in balance, as seen from the following thought experiment.

---

[1] See pages 169–70 in the appendix on centers of mass.
[2] See page 161 in the appendix. Here I am referring only to the horizontal forces.

29

First, according to Archimedes, we can replace the duck with the water it displaces without any effect on the balance of the dish. And now removing the duck amounts to slowly sucking out some of the water. The water has the time to redistribute itself, and so we can just think of removing a layer of water of constant thickness; this will not affect the balance.

Question. Is the answer to the problem the same if the dish is not symmetric, as in Figure 4.2b?

Answer. The balance may be lost. In Figure 4.2b, if we remove water above the dotted line, we will shift the center of mass of the water to the left, and the dish will tip to the left.

4.2 The Tub Problem—In More Depth

Do not read this puzzle before you read the preceding one.

A puzzle. Figure 4.3 shows a toy boat from the preceding puzzle, but pulled down by a cable attached to a wheel rolling along an underwater rail affixed to the tub. Just as in the preceding problem, the boat starts at rest, travels to the other end of the tub, and stops; everything comes to rest. And the question is the same: Will the tub shift? If so, in which direction?

Some ideas. In the preceding puzzle, where the cable was absent, we established that the tub will stay in place. Now how could a vertical force of the cable possibly affect

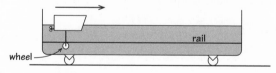

Figure 4.3. Which way will the tub shift with the cable pulling the boat down?

horizontal motion of the tub and the boat? Naturally the tub will remain fixed, just as in the preceding problem.

Is this conclusion correct, and if not, where is the mistake in the reasoning?

The (correct) answer. The tub moves in the same direction as the boat. The explanation is similar to the one for the space shuttle puzzle on page 5—this is in effect the same problem. Since the boat is being pulled down, the mass of displaced water is greater than the boat's mass: the boat takes up a lot of underwater volume but contains little mass, just like the helium balloon in the capsule on page 6. In effect, the less massive boat swaps places with the more massive water after the boat's trip. In short, from the tub's point of view, the center of mass moves left. But actually the center of mass is fixed relative to the ground—so it is the tub that moves right.

Putting it differently, the linear momentum of the entire system is conserved, so if the overall mass inside the tub moves left (relative to the tub), the tub itself must move right in order to conserve zero linear momentum of the entire system.

Question. Where is the mistake in my "proof" that the tub won't move?

Answer. The mistake is in the claim that the cable does not affect horizontal forces on the tub. Indeed, when the boat moves it displaces the water, and the deeper draft of the boat results in greater displaced water volume. And this water interacts with the walls of the tub. So the vertical pull of the cable does have a horizontal effect.

Negative masses. The boat's mass is less than the mass of water it displaces, since the boat is being pulled down. We could thus say that the boat's mass is negative in comparison to the mass of the water it displaces. It is this negative sign that is responsible for the counterintuitive result.

4.3 How to Lose Weight in a Fraction of a Second

A puzzle. You are standing on the bathroom scale. Can you make the scale show less weight (without leaning on anything, or dropping any clothes)?

Answer. This was a trick question: the answer is yes, but only for a short time. All you have to do is to bend your knees. If you bend your knees *really* fast, your feet will lose contact with the scales altogether, and the scales will show zero weight. With a slower motion, the effect will be less dramatic, but the weight decrease will still be there. Of course, soon you will have to stop accelerating down, and then the scales will show a greater weight, until you stop moving.

All this is explained by Newton's law[3]

$$ma = F,$$

where a is the acceleration and $F = S - W$, where S is the force with which scales push me up, and W is my weight. The scales always show the force S. Now if I bend my knees, I accelerate down: $a < 0$, and so

$$ma = S - W < 0,$$

or $S < W$; scales show less than my weight. If I stand still, then $a = 0$ and $S - W = 0$; the scales tell the sad truth. And as I jump up, making $a > 0$, we get $S - W > 0$, or the scales show more than my weight.

---

[3] Stated on page 172 in the appendix.

The following question requires a little calculus.

Problem. Show that no matter how I jump on the scales, the average indication of the scales will approach my true weight, provided I wait long enough.

Solution. Let T be a long waiting time. Integrating $ma(t) = S(t) - W$, we get

$$\int_0^T ma(t)\, dt = \int_0^T (S(t) - W)\, dt.$$

Now $\int_0^T a(t)\, dt = \int_0^T v'(t)\, dt = v(T) - v(0)$ by the fundamental theorem of calculus (page 184); the above gives

$$m(v(T) - v(0)) = \int_0^T S(t)\, dt - WT.$$

Dividing by T we get

$$\frac{m}{T}(v(T) - v(0)) = \underbrace{\frac{1}{T}\int_0^T S(t)\, dt}_{\text{average of } S} - W.$$

As $T \to \infty$ (assuming I live that long), the left-hand side approaches zero, since v is bounded due to my human limitations. Thus the right-hand side approaches zero as well, and thus the average $(1/T)\int_0^T S(t)\, dt$ approaches W, as claimed.

4.4 An Underwater Balloon

Question. An air-filled balloon is held under water by a string tied to the bottom of a jar. The jar stands on a scale. The string suddenly breaks. What happens to the reading of the scale just after the break?

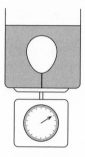

Figure 4.4. Just after the string breaks, will the reading of the scale increase, decrease, or stay the same?

True or false? Before the string breaks, it pulls the bottom of the container upward. Once the string breaks, the upward pull on the container disappears and the container then feels heavier. So the scale will show increased weight right after the string breaks.

The truth. Actually, the opposite is true: the scale will initially show a *decreased weight*; the jar will feel lighter. To understand why, let's see what happens to the center of mass of the system (the jar with all its content). After the string breaks, the balloon accelerates up, and the water's center of mass accelerates down. But water is a lot denser than the balloon, and so the net result is the downward acceleration of the center of mass of the contents of the jar. And this means less force on the scale—just as in the preceding problem when I stood on a bathroom scale and suddenly bent my knees, causing the scale to show a decreased weight. More formally, by Newton's second law (page 172), the acceleration a of the center of mass is due to all the forces on the jar, of which there are only two (reaction from the scale and weight):

$$reaction - weight = ma.$$

Now when the string breaks, the initial acceleration is down, that is, $a<0$, implying that *reaction* − *weight*<0, or *reaction* < *weight*, that is, the reaction force (which is what the scale shows) is less than the weight.

Where is the mistake in the initial reasoning? The mistake was in not telling the whole truth. Unmentioned was the force of the *water* on the jar. In fact, the moment the string is cut, the water starts dropping, causing the hydrostatic pressure on the bottom of the jar to lessen, making the jar feel lighter. True, cutting the string tries to make the jar heavier, but not as much as the lesser pressure makes it feel lighter. Our center-of-mass argument shows that the lightening-up effect wins.

A moral lesson. This problem shows how deceiving looks can be. The balloon is visible but very light. Although the water is more massive, and thus vastly more important, I overlooked its motion, perhaps because the water is transparent. We misdirected our attention to the visible but unimportant away from the invisible but important. Physics imitates life, where lightweights sometimes attract more attention than they deserve.

4.5 A Scuba Puzzle

Question. A submerged scuba diver, slightly buoyant, maintains his position somewhere in the middle of a rectangular tank, by working his fins against buoyancy. Next to this tank, we have an identical tank with no diver, filled to the same level.[4] Which tank weighs more?

---

[4] In other words, the volume of water + diver in the first tank is the same as the volume of water in the second tank.

The paradox. Here are two conflicting arguments. (A): On the one hand, the first tank is clearly lighter since the volume of its contents is the same as that of the second tank, while the diver's density is less than that of the water. (B): On the other hand, since the tanks have the same depth, the water pressure at their bottoms is the same as well. So the tanks must weigh the same.

Where is the mistake?

Answer. Argument (A) is correct; it is hard to argue with algebra. So let's find the mistake in (B). The buoyant diver has to move his fins to stay submerged; in the moving water, the pressure need not be the same as in still water at the same depth. In fact, even the term "depth" is no longer precisely defined, since the water surface is not flat when the diver moves his fins. A small mound forms above the diver because he propels the water upward in order to stay submerged. From argument (A) we conclude that the average water pressure at the bottom of the tank with the diver is less than that for the other tank.

Here is an interesting question which I leave to the reader:

Problem. Consider a helicopter hovering above the water. The downwash creates a shallow indentation in the water surface. Does Archimedes' law hold here? That is, does the weight of the helicopter roughly equal the weight of displaced water? Assume that the motion of the air and of the water is steady.

4.6 A Weight Puzzle

Two jars in Figure 4.5 are filled with water to the same level. The bottoms of both jars have the same diameter.

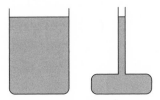

Figure 4.5. Does water in the left container press on the bottom with greater force?

Question. Is it correct to say that the force of water on the bottom of the left container is greater than the one on the right?

Answer. It is false: both bottoms feel the same force. The reason is that the pressure—the force per unit area—depends only on the depth and not on the container's shape: whether in a cup, or in a lake, the pressure at the same depth is the same! Since the bottoms of both containers are at the same depth, the pressures there are equal. And since the bottoms have equal areas, the forces on them are equal as well.

Question. What is wrong with this logic: "Since the water in the right container weighs less, it presses on the bottom with lesser force"?

Answer. Figure 4.6 explains that the bottom of the bottle feels force greater than just the weight of the water. The excess force equals the downward force from the "ceiling." Indeed, by Newton's first law the forces on the water are in balance: the down forces balance the up force:

$$weight + force\ down = force\ up,$$

so that $weight < force\ up$. The weight of the water is less than the upward force from the bottom.

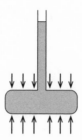

Figure 4.6. The force exerted by water on the bottom depends only on the depth below the free surface and on the area of the bottom, but not on the weight of the water.

The neck in the bottle may be thin, but it generates the same pressure as a thick neck would!

🎇 5 🎇

FLOWS AND JETS

5.1 Bernoulli's Law and Water Guns

Question. Imagine shooting water from a syringe by pushing the piston. With Newton's first law in mind (motion is steady if no force is applied), I ask: Does it take any force to move the piston with a constant speed—assuming a perfectly frictionless piston and perfectly nonviscous water? In other words, once I push the piston to give it some speed and let it go, will it continue at the same speed by inertia?

Answer. A force *is* required to push the piston with constant speed—even in a perfectly frictionless world. Newton's first law of steady motion by inertia doesn't apply here because *some parts of fluid do accelerate*—namely, those nearing the exit from the cylinder. Figure 5.2 shows this in more detail.

Figure 5.1. Will the piston move steadily by inertia if there is no viscosity and no friction?

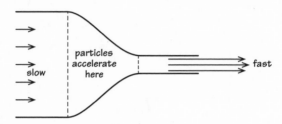

Figure 5.2. There are always particles of water that speed up. Pressure difference is needed for this speed-up.

This speed-up signals that the particles of fluid are being "pushed from behind," that is, that the pressure is higher behind a particle than in front of it. This is the simple gist of *Bernoulli's law,* often stated as "the pressure is lower where the speed is higher." This may sound a bit misleading, as if the higher speed causes lower pressure. It's rather the other way around: "the fluid accelerates in the direction of lower pressure." So if the pressure downstream is lower, then the speed there is higher.

This is both the statement of Bernoulli's law and its explanation. So Bernoulli's law is a special case of Newton's second law.

Bernoulli's law suggests an analogy with a falling rock: the lower the rock's level/potential energy, the faster it moves. In fact, Bernoulli's law can also be viewed as a special case of conservation of energy.

Problem. What force is required to push the piston with a constant speed v? The area of the piston is A, and the area of the exit tube is a.

Solution. When we use force F to move the piston a distance D, we do work $W = FD$. This work is spent entirely on increasing the kinetic energy of the water (this is where we assumed no friction or viscosity):

$$F \cdot D = \frac{m v_{\text{exit}}^2}{2} - \frac{m v^2}{2}, \qquad (5.1)$$

where m is the mass of the water expelled. It remains to solve for F, and to express everything in terms of v, a, and A. To that end we note first that $m = \rho A D$, where ρ is the density of the water and A is the area of the piston's face. Also, since the water is incompressible, the volume pushed out by the piston per second (vA) equals the volume that left the syringe: $vA = v_{\text{exit}}a$, where a is the cross-sectional area of the exit tube. Thus

$$v_{\text{exit}} = \frac{A}{a} v,$$

which we could have guessed directly. Substituting all into (5.1), we obtain

$$F = \frac{1}{2} \rho A v^2 \left(\frac{A}{a} - 1 \right) \qquad (5.2)$$

Two interesting consequences follow. Let's fix the piston's speed v and see what happens for different area ratios A/a.

1. For a very narrow exit tube: If A/a is large, the required force F is large, according to (5.2). Pushing the water out through a narrow hole is hard, but not, as one may think, because of viscosity. Rather, it is hard because we constantly accelerate particles. The work we do does not go into heat (kinetic energy of disordered motion) but rather into the kinetic energy of the ordered motion of the ejected water.

2. If the tube widens, rather than narrows: $A/a < 1$ as in Figure 5.3, then $F < 0$, according to (5.2). This means that we have to *pull* the piston to maintain constant speed! Again, this is clear directly without

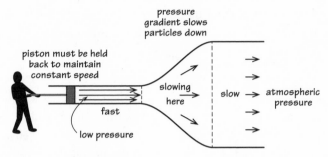

Figure 5.3. Maintaining constant speed in a widening tube requires a steady pull against the motion.

any formulas, since the water comes out with smaller speed than it has inside the cylinder. This means that we must slow the particles of water by pulling the piston.

5.2 Sucking on a Straw and the Irreversibility of Time

Question. When using a straw, does it take greater effort to suck the water in or to blow it out, Figure 5.4? The question assumes that water in the straw moves with constant speed, and that gravity plays no role.

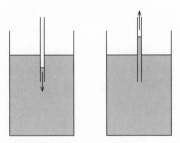

Figure 5.4. Does it take the same pressure in the two straws to maintain constant speed?

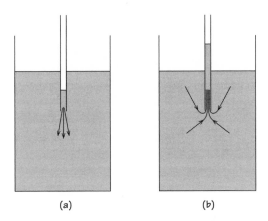

Figure 5.5. Is it harder to suck the water in than it is to blow it out (all at constant speed)?

Answer. Sucking is harder; Figure 5.5 explains why. Under the suction, the water enters the straw from all directions, as in the part (b) of the figure. A typical water particle gains a lot of speed as it enters the intake. Suction is needed to provide for this acceleration.[1] Putting it differently, we must spend energy to accelerate the fluid, and this requires that we pull hard.

By contrast, Figure 5.5a shows that the ejected water comes out as a jet. Since the jet widens gradually, the pressure change along the stream is also gradual.

The arrow of time. One might think that by reversing the direction of flow in the straw we would simply reverse the direction of motion throughout the water; we know from childhood that this is not the case. Blowing out a candle is easy, but it is impossible to extinguish it by suction (certainly not from a safe distance, without burning one's lips; I would

---

[1] This is precisely the Bernoulli effect from page 39.

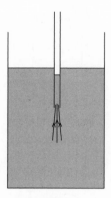

Figure 5.6. Flow that starts as an incoming jet is unstable and will soon begin to look like the one in Figure 5.5.

be interested to hear from anyone who succeeded (but not from their lawyer)). This mystery of preferred direction of time has fascinated scientists for many years. The mystery here lies in the following seeming contradiction. Speaking just of classical mechanics, Newton's laws are time-reversible. And yet, if we consider a large collection of classical particles, such as the ideal gas, the time reversibility seems to be lost. Resolution of this seeming contradiction lies in the fact that the flow of Figure 5.6 (the time reversal of Figure 5.5a) is theoretically possible, but extremely unstable: even if we started the fluid as in Figure 5.6, it would almost immediately disintegrate and will begin to look like in Figure 5.5b.

5.3 Bernoulli's Law and Moving Around in a Space Shuttle

Question. Imagine yourself floating motionless in the middle of a cabin in a space shuttle. Having had enough rest, you now want to reach a wall. You could throw something

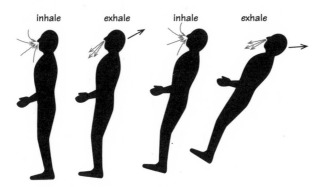

Figure 5.7. A person in weightlessness is propelled by his breath.

to start moving in the opposite direction—say, your shoe or your belt[2]—but if you are not allowed to throw things, can you still reach the wall?

Answer. Just keep breathing. When inhaling, you pull the air in from many directions; when exhaling, the air comes out as a jet, just as in Figure 5.7. The net result of one inhalation–exhalation cycle is to push air in the direction of that jet; you will then be pushed in the opposite direction. You will thus be acting as a very inefficient squid.[3] A mouth-breathing person will move slower (no surprise here) since the air is expelled through the mouth with lesser speed than it is through the nose.

5.4 A Sprinker Puzzle

Question. Our sprinkler is an S-shaped pipe spinning on the pivot P; water is supplied through the pivot, as sketched in

---

[2] Does one really need these in weightlessness? One cannot walk, and the pants don't fall down.

[3] Squid propel themselves on the same principle, except that they shoot the water out of the rear.

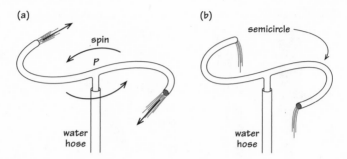

Figure 5.8. In which direction does the jet leave the nozzle?

Figure 5.8, and the jet force makes the sprinkler spin. What is the direction of the exiting water relative to the ground observer? Assume a frictionless pivot and steady spinning speed.

Answer. The water leaves the nozzle in a purely radial direction, that is, directly away from P—the direction shown in Figure 5.8a is wrong! The tangential component of the velocity of the water is zero, as in Figure 5.9.

An explanation. The water supplied by the hose has no spin around the vertical axis. And the only thing that could impart spin to this water is the rotational friction at the pivot—but that's zero by assumption. So water exits the nozzles with the same spin it entered: zero. This explains the purely radial velocity.[4]

A follow-up puzzle. A designer thought of changing the sprinkler as shown in Figure 5.8b, by shaping each arm as a semicircle. Is this a good design?

---

[4] This explanation can be made more precise simply by replacing the vague term "spin" by the precise term "angular momentum" and by referring to the fact that the angular momentum does not change due to the absence of torque.

Solution. The idea is terrible: the water will exit the nozzles with zero ground speed, simply plopping down!

Here is why. The velocity of exiting water is purely radial, as we had established. But the radial velocity at the exit must be zero because the arm is a semicircle. This is a strange sprinkler: the water comes in with some speed, but it leaves with zero speed.

A puzzle. The water in the supplying pipe has kinetic energy. The outgoing water is deposited with zero speed, thus having zero kinetic energy. What happened to the kinetic energy?

Solution. In answer to the last question, the sprinkler sucks—literally. This is not a cry of frustration but a statement of physical fact, which I now explain. By saying that the sprinkler sucks I mean that the pressure inside

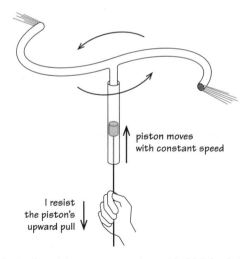

piston moves
with constant speed

I resist
the piston's
upward pull

Figure 5.9. As the piston moves up, I must hold it back to keep the speed constant.

the supplying hose is negative,[5] Figure 5.9. This happens because the water in the spinning pipe is thrown outward by centrifugal effect, thus creating suction.

A liquid "whip." How would the sprinkler behave if the piston in Figure 5.9 were not held back? The spinning arms provide the centrifugal suction, which speeds up the water in the pipes. The sprinkler will spin faster and faster, until all the water is gone. This is very similar, strange as it may sound, to the cracking of a whip. Snapping a whip sends a wave along the rope. As this wave travels toward the tip, it shortens, so that the same energy is concentrated in the ever-shorter wave. If done right, this concentration can be so high that the tip can even exceed the speed of sound. A similar energy concentration, although less dramatic, happens in our thought experiment with the sprinkler.

5.5 Ejecting Water Fast but with Zero Speed?

Question. Figure 5.10 shows a container with a rubber hose attached. Is it possible to make water exit from the hose with zero speed?[6]

Answer. All we have to do is move the end of the hose with velocity opposite to that of exiting water; the water then comes out at zero speed. This is similar to throwing an apple from a moving car backward, with the speed equal to that of the car. The apple will have zero ground speed at the moment of its release.

---

[5] In other words, below the atmospheric pressure.

[6] The speed is measured relative to the ground observer.

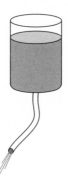

Figure 5.10. Can one eject water from the tank so that the ejected water has (nearly) zero speed?

In fact, the dysfunctional sprinkler in Figure 5.8b will do the job of ejecting water with zero speed if we connect it to the container.

The dysfunctional sprinkler used to empty the tank has a remarkable advantage: it offers a way to transfer liquid from one container to another very fast, without splashing and yet with no pump. Indeed, since the sprinkler creates suction (as explained on page 47), it works as a pump!

5.6 A Pouring Water Puzzle

The following question occurred to me while washing the dishes and watching the water coming out of the faucet strike the bottom of the sink.

Question. Water pours steadily out of a jar,[7] striking a disk attached to scales, and scattering off sideways as shown in Figure 5.11. The scales register a certain force due to the

---

[7] I replaced the faucet with a jar to avoid dealing with the pipe, which starts at some indefinite place.

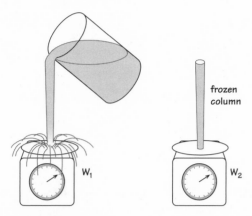

frozen
column

W₁

W₂

Figure 5.11. How does the impact force from the jet compare to the weight of the water column?

impact of the jet. Which is greater: the jet's impact force or the weight of the airborne column of water? Or are they about the same?

The air resistance, surface tension, and other relatively minor distractions are to be ignored. The vertical speed of water at the mouth of the jar and inside the jar is to be neglected.

Answer. Only a small amount of water at a time hits the plate, suggesting, maybe, that the impact force is less than the weight of the whole column. But actually the two forces are roughly equal!

An informal explanation. The linear momentum[8] of the entire water is constant while the jet is pouring steadily— indeed, the linear momentum of the jet is constant since

---

[8] Linear momentum is explained in the appendix, page 171; I am speaking here of the linear momentum in the vertical direction.

the jet remains the same during pouring, while the linear momentum of all the other water is zero.

As explained at the end of Section A.4, the constancy of the linear momentum implies that the sum of forces on water is zero: $W - R = 0$, where W is the weight and R is the sum of upward reaction forces from the jar, from the scale, and from the ground. The relation $W = R$ amounts to

$$W_{jar} + W_{column} + W_{ground} = R_{jar} + R_{scales} + R_{ground}. \quad (5.3)$$

But $W_{jar} = R_{jar}$ and $W_{ground} = R_{ground}$ since the water in the jar and on the ground is essentially at rest. Canceling the terms in (5.3), we get

$$W_{column} = R_{scales},$$

as claimed.

5.7 A Stirring Paradox

Background on fluids. All the background we need for this puzzle consists of the following sentence, which is deciphered in the next paragraph: "The vorticity of the nonviscous fluid remains zero if it is zero initially." This is a special case of *Kelvin's theorem*.[9]

What is vorticity? As the term suggests, vorticity measures the rotation, in the following precise sense, which I explain only for *two-dimensional fluids*. Let me inject two short

---

[9] The full statement and proof of this theorem can be found in, e.g., Batchelor. Speaking of two-dimensional fluids, here, I note that an alternative and (I think) more transparent proof is almost immediate from the fact that (1) a *circular* blob of fluid feels zero torque (around its center) due to the lack of viscosity, and (2) the area of the blob does not change as it is carried by the fluid (the blob need not remain round; it is only assumed to be round at one instant).

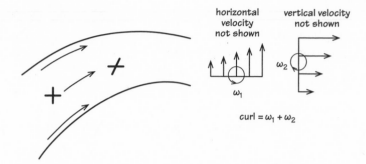

Figure 5.12. Vorticity of fluid at a point is the sum of angular velocities of two infinitesimal dashes at the instant when they are perpendicular to each other.

dashes of dye in the fluid, as shown in Figure 5.12. The dashes may turn as they are carried along (and also stretch, but I don't care about that); let me record the angular velocities ω_1 and ω_2 of these dashes at the initial moment, when they are still perpendicular. *The vorticity of the fluid at the point p is, by the definition, the sum $\omega_1 + \omega_2$.* For the fluid rotating as a rigid body, the vorticity is simply twice the angular velocity of its rigid rotation.

The turning fluid paradox. Figure 5.13 shows ideal fluid filling a ring with a piston inside.[10] All is at rest. We then push the piston all the way around the ring and stop. Throughout the process the vorticity must remain zero according to Kelvin. But how can this be, given the fact that the water turns around?

Solution. As the piston travels counterclockwise, the water does not move as a rigid object; if it did, vorticity indeed

---

[10] All this discussion is in dimension two. Alternatively, the reader can think of Figure 5.13 as the cross section of a three-dimensional object.

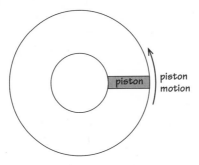

Figure 5.13. How can fluid rotate while keeping zero angular velocity (more precisely, vorticity)?

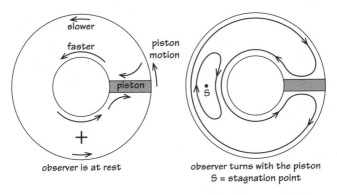

Figure 5.14. (a) The observer is stationary; the flow is faster on the inside track. (b) The view for the observer who rotates together with the piston.

would not have been zero. The water keeps its zero vorticity by moving faster near the inner circle than it does near the outer one, as shown in Figure 5.14a. The fluid performs two simultaneous motions: it (1) turns counterclockwise with the piston and (2) it circulates clockwise. The flow looks especially simple in the frame rotating with the piston, Figure 5.14b.

Question. Is there a point in the fluid that comes back after one revolution of the piston?

Answer. A point P diametrically opposite the piston returns to its original position. In fact, this point stays fixed relative to the piston throughout the revolution.[11]

5.8 An Inkjet Printer Question

Inkjet printers work by shooting thin jets of ink into paper.

Question. Water (or ink) is ejected from a thin tube and breaks into droplets, due to surface tension (Figure 5.15). Do the droplets fly with the same speed v as the water in the tube? Ignore gravity and air resistance.

Answer. The droplets move slower than the ink in the tube. Surface tension causes the inkjet J in the figure to behave a bit like a rubber band. This tension pulls the tip T to the left, toward the mouth of the tube, slowing the tip down. Later this tip breaks off and becomes a droplet, now slower than the ink in the tube.

Figure 5.15. Do the droplets fly with the same speed as the emerging jet? Air resistance is to be neglected.

---

[11] The existence of the "returning point" follows from the Bohl-Brouwer fixed-point theorem, whose statement and proof can be found in any undergraduate topology text, e.g., in J. Munkres, *Topology* (Upper Saddle River, NJ: Prentice-Hall, 2000).

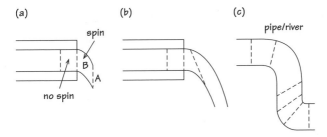

Figure 5.16. Because the layers of water begin to drop as they come out of the pipe while those in the pipe are not yet falling, vorticity is created, or is it?

5.9 A Vorticity Paradox

Question. Water comes out of a horizontal pipe as shown in Figure 5.16. As the layer A comes out, it starts falling; the layer B is still in the pipe, so it doesn't fall yet. This means a shear flow, in a clockwise direction, that is, a clockwise vorticity.[12] When emerging from the pipe, water changes its vorticity from zero. But this contradicts Kelvin's theorem (stated above), according to which vorticity stays constant. Where is the mistake in the argument?

Answer. My mistake is revealed in Figure 5.16b: after emerging from the pipe, the layers do not remain vertical but rather tilt as shown. This tilt cancels the rotation mentioned in my original argument.

Flow in the pipe. A similar "counterrotation" is observed in water going around a bend in a pipe or a river; Figure 5.16c

---

[12] Vorticity was defined on page 52.

shows how a line of dye travels through a bent pipe.[13] When the pipe turns right, the line of dye rotates left in such a way as to preserve zero vorticity. In the second bend of the pipe the opposite happens: the pipe turns left, the line rotates right.

---

[13] I am assuming that the fluid is perfectly nonviscous, so that the particles are not slower near the walls.

❈ 6 ❈

MOVING EXPERIENCES: BIKES, GYMNASTICS, ROCKETS

6.1 How Do Swings Work?

Question. Most things in life are easier said than done. But some are the other way around—easier done than said. Rocking on swings is an example. Exactly how does a child pass the energy of his muscles to swinging? The answer is not that obvious.[1]

Answer (the anatomy of a resonance). Imagine yourself rocking back and forth on swings. You feel the greatest g-force when zooming past the bottom of your path. By the same token, the least g-force is near the highest points of your trajectory.[2] Imagine now holding a weight in your hand, resting it your lap, Figure 6.1. When you are passing the bottom of your trajectory, raise the weight to your shoulder; keep it there until you are near the top of the trajectory and once there quickly lower it back. Repeat: lift when passing

---

[1] A little boy asked: "Grandpa, do you sleep with your beard over or under the blanket?" That night, the sleepless grandpa was lying on his back, trying his beard this way or that and cursing. Neither way felt right any more.

[2] There is a "double whammy" responsible for the lesser g-force at the top of the trajectory: (1) the centrifugal force is smaller, and (2) gravitational force's component is smaller.

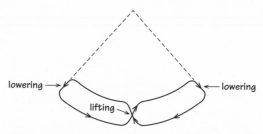

Figure 6.1. This path of the center of mass will feed energy into swinging motion.

the bottom, lower near the top, and so on. This action will make each subsequent swing higher. Why? Because you do positive work: by lifting the heavier weight you spend more energy than you get back by lowering the lighter weight. The energy mismatch goes into increasing the amplitude of oscillations.

Of course, you don't have to bring a weight: you can use your head (literally), torso, or legs. In fact, this is exactly what kids on swings do: they straighten the knees and sit up at the bottom of the trajectory (lifting weight) and bend the knees and lean back (lowering weight) when near the top of trajectory.

As a child I could do all this, but could not explain it. Now, it may be the other way around.

6.2 The Rising Energy Cost

Question. A rock falls with constant acceleration (we ignore the air resistance). Figure 6.2 shows the rock's speed after each subsequent meter traveled. Why do the gains in speed decrease with each subsequent meter?

Answer. As the rock accelerates, it spends less and less time passing each subsequent meter, and consequently has less and less time to gain speed while passing each meter.

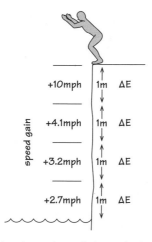

Figure 6.2. The farther down the rock drops, the less speed it gains per meter traveled.

Here is a more formal explanation. If the rock is released at the height h above ground, its potential energy[3] is mgh (and zero kinetic energy since it is released at zero speed). Just before hitting the ground, all the energy is kinetic: $mv^2/2$. Equating the two expressions and canceling m gives

$$\frac{v^2}{2} = gh,$$

or $v = \sqrt{2gh}$. Now the graph of v versus h is a parabola lying on its side; the slope of this parabola is decreasing. And so equal increments in h give decreasing increments of v for larger h.

Question. How do we know that the weight of the object of mass m is mg?

---

[3] Potential energy is, by the definition, the work required to lift the mass to height h. This work = force · distance = weight · h = mgh, since weight = mg.

Answer. By the definition, g is the acceleration caused by the gravitational force W upon the freely falling mass, that is, by the weight. Thus, by Newton's second law ($F = ma$; see page 161), we have $W = mg$.

6.3 A Gymnast Doing Giants and a Hamster in a Wheel

The *giants* on a high bar is a basic gymnastics move in which the gymnast starts in a handstand, swings downward past the ground, and swings back up past a handstand again, and so on (Figure 6.3a).

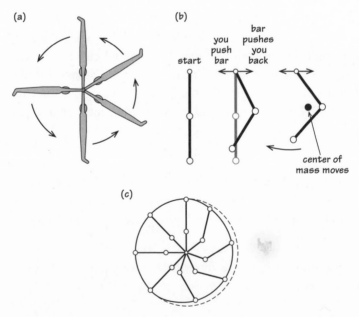

Figure 6.3. (a) Doing giants on a high bar. (b) How to activate gravitational torque. (c) How to feed energy into giants.

Question. The gymnast starts out by hanging still from a high bar. There is no friction between her hands and the bar: her grips are perfectly slippery, so that no torque can be created by moving her wrists. Can she start giants without friction?

Ignore the flexing of the bar, the air resistance, and other relatively minor things.

Many physicists and mathematicians give an answer which usually goes like this: "Without friction there is no torque; without torque, no angular momentum can be created, and thus no rotation can be created. So the gymnast cannot start giants without friction."[4]

This is a false argument. Luckily for them, the young gymnasts who do giants don't know enough physics to be stopped by such logic. They don't know about the torque and the angular momentum—their age may be less than the number of years it takes to get a Ph.D. Sometimes, knowing less can be an advantage.

Where is the error? The zero-torque premise is false: *gravity* may exert torque. True, this torque is zero when the gymnast just hangs still. But by bending her body she can bring the torque of gravity into play; here is how.

Imagine yourself hanging on the bar, Figure 6.3b. Start bending your body at the waist, as if you were trying to touch your toes. As you tense up your stomach muscles, your hands want to go forward, thus pushing on the bar forward (left in the figure). The bar pushes you right, by Newton's third law (action equals reaction). So your center of mass shifts right as well, by Newton's second law. Now your center of mass is no longer directly under the bar, so

---

[4] Angular momentum is described in the appendix.

that gravity now exerts torque, trying to swing your center of mass left, as it would a pendulum. To summarize, by bending your body you enabled gravity to exert the torque.

Question. Once already doing giants, what (in principle) should a gymnast do to speed them up?

Answer. The principle is exactly the same as that for the swings: do net positive work. To that end, bring your center of mass closer to the bar when it is harder to do (near the bottom) and move it farther from the bar when it is easier, near the top.

A crude parody of this is shown in Figure 6.3c.

An unbalanced wheel. Here is an alternative way to understand the gymnast's behavior. Figure 6.3c shows the path of her center of mass. Imagine now smearing the mass over the path, and think of the path as the rim of a wheel whose axle is the bar. Since the wheel is offset to the left, gravity will apply counterclockwise torque to this wheel. Such an offset can be maintained by adjusting the spokes of the wheel in real time, thus causing a persistent gravitational torque to act on the wheel. This torque is what allows the gymnast to gain speed and to compensate for friction.

In fact, this is similar to what a hamster in a wheel does: he keeps the center of mass offset to one side of the axle, thus maintaining the torque that spins the wheel, although the hamster probably does not think of this in such terms.

Problem. Changing the lengths of spokes in the adjustable wheel requires work. Explain in more detail how to change the lengths of spokes to maintain the offset of the rim.

Note that the tension of the shortening spokes is greater (on average) than the tension of the lengthening spokes. Recall that the length of the spoke is like the distance from the gymnast's center of mass to the bar.

6.4 Controlling a Car on Ice

I learned this problem from Andy Ruina of Cornell University.

Problem. Imagine driving a car in a straight line on a large field of ice. You hit the brakes. Of course, it's best not to have the wheels lock, but if it's going to happen, which would you choose to have locked: front or rear? The goal is to stop while staying in a straight path, with no spinout.

Solution. Surprisingly, it's better for the front wheels to lock.[5] With front wheels locked, the car will keep going straight. By contrast, if the rear wheels lock instead, the car will turn around and will travel with back end forward (assuming the steering wheel is kept fixed) until it stops. A cyclist may notice a similar effect: if the rear wheel locks, it will slide out sideways (assuming that the front wheel is not locked).

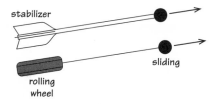

Figure 6.4. Just like the feathers keep an arrow straight, the rolling rear wheels keep the car (with locked front wheels) going straight.

---

[5] You won't be able to steer, but the goal here is only to stay in a straight path.

An explanation. Analogy with an arrow is illuminating. The arrow flies in a straight path because the feathers keep its tail from sideslipping (Figure 6.4). Similarly, in a car, with front wheels locked, the rolling rear wheels act just like the stabilizing feathers. And thus the car is stable when these rolling wheels are in the rear and is unstable when they are in front.

At a late hour one snowy night I tried this experiment in my car on an empty snow-covered parking lot. By locking the rear wheels (using the parking brake), I could easily turn the car 180°.

6.5 How Does a Biker Turn?

Problem. A biker, going straight, wants to quickly turn left. What does he do with the handlebars?

Solution. To turn left, he needs to first get the bike to lean left to compensate for the centrifugal force during the turn. To create the left lean, he momentarily moves the handlebars to the right, causing the wheels to ride to the right from under him while his body goes straight by inertia (Figure 6.5). Having thus achieved the desired lean to the left, he now turns the handlebars to the left and executes the desired turn. If you try this experiment on yourself (as I did), you will notice how your hands subconsciously do the initial counterturn. Our subconscious reflexes are great teachers of mechanics.

There is an additional effect on a fast-moving bike with massive tires: the gyroscopic effect of the front tire helps to create a lean. By pushing the handlebars to the right you create a very noticeable lean to the left.

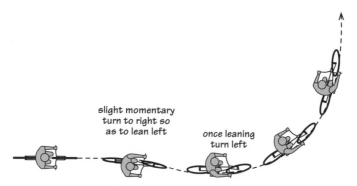

slight momentary
turn to right so
as to lean left
once leaning
turn left

Figure 6.5. To turn left, the biker needs a small initial turn of the handlebars to the right in order to create a lean.

6.6 Speeding Up by Leaning

Question. Can you change the bike's speed by steering only? No pedaling or moving the body is allowed.

Answer. If you start with the bike going in a straight line, just by leaning into a turn you will automatically increase your speed.[6] The reason: by leaning you decrease your potential energy. The kinetic energy must therefore increase, and so therefore must the speed.[7]

Does the gain in speed depend on the starting speed (assuming the same leaning angle)? Surprisingly, the gain is greater at lower speeds.[8] The same lean angle while traveling at 1 mph will gain you more speed than at 10 mph. Here is an explanation. When I lean, thereby lowering my center of

---

[6] What a rider does to lean into a turn was described in the preceding problem.

[7] Strictly speaking, some kinetic energy goes into the kinetic energy of rotation—after all, the biker is now turning. One can show, however, enough energy is left to cause an increase in the speed.

[8] This is related to the fact that jumping from four times the height results in a gain of only twice the speed.

mass by h, I gain kinetic energy equal to the lost potential energy:

$$\frac{mV^2}{2} - \frac{mv^2}{2} = mgh,$$

where V is my new speed and v is the old speed.[9] Simplifying, we get

$$V^2 - v^2 = 2gh, \quad \text{or} \quad V - v = \frac{2gh}{v+V} < \frac{gh}{v}.$$

This shows that the speed gain $V - v$ is smaller for larger initial speed v.

6.7 Can One Gain Speed on a Bike by Body Motion Only?

By merely leaning into a turn, a biker gains speed (as just explained)—however, this is only a small, one-time increase that cannot be done in a cyclic fashion to maintain speed indefinitely.

Problem. Can a bicyclist increase his speed indefinitely (theoretically speaking) without pedaling, only by moving his body?

Just to eliminate the possible loopholes, there is no wind, one cannot use any engines, and so on.

A hint. A wheel rolling on the ground is like a skate on ice in this one respect: both move very easily in the direction they point in, and neither wants to move sideways.

---

[9] When going in a circle, I also acquire some kinetic energy of rotation, which I neglect.

Solution. I start out coasting in a straight line, sitting upright. My goal is to end up in the same state but with a greater speed. I can achieve this in three steps:

1. Lean down over the handlebars, thus lowering my center of mass.

2. Turn into a tight circle; once there, I straighten up, thus raising my center of mass.

3. Start going straight again.

Why do these actions increase the speed? Note that the g-force is greater when going in a circle, due to the extra centrifugal effect.[10] I do more work straightening up against this greater g-force than what I get back by lowering my center of mass.[11] The difference of these energies goes into the gain of my kinetic energy. The same principle is used by a child to put energy into swings or by a gymnast doing giants.

An alternative explanation. Another way to explain this gain of speed is by using the conservation of angular momentum. When going in a circle, my angular momentum around the center of the circle is constant (since the torque upon me around the center of the circle is zero (see page 176 on the conservation of angular momentum). By the act of sitting upright, I move my center of mass closer to the center of the circle. My velocity must therefore increase in order to preserve the angular momentum.

---

[10] See the next problem.

[11] When saying that "I get back" energy I imagine using gravitational energy to charge a battery, like a hybrid car does when braking.

6.8 Gaining Weight on a Motorbike

Question. You are on a motorbike, going in a perfect circle at a constant speed. You are leaning steadily into the turn. What g-force do you experience? In other words, what is your apparent weight?

Answer. Figure 6.6 shows two forces felt by the bike: (1) the reaction force R of the ground, and (2) his weight W. The reaction force R is the perceived weight. Now the resultant force of these two forces is the centripetal force[12] that makes the bike travel in a circle. Therefore, this resultant is pointed at the center of the circle, that is, it is horizontal, as shown in Figure 6.6. The triangle ABC is therefore a right one, and its angle A is the same as the biker's lean angle θ. From $\triangle ABC$ we have

$$R = \frac{W}{\cos \theta}. \tag{6.1}$$

Since $\cos \theta < 1$, we have $R > W$; one always feels an increased g-force in a steady turn. For $\theta = 30°$ this amounts to a 15% increase in perceived weight. If you can manage to sustain the lean of $45°$, the relative increase in weight is 41%. A 180-pound person will feel like 254 pounds. And if you lean at $60°$ to the vertical—a huge lean—you double your perceived weight. The same formula works for an airplane in a steady turn. For example, to sustain the steady force of $2g$ the pilot must tilt the plane at $60°$ (so that the wings form $30°$ with the vertical). This also shows that when you are going around a ramp in a car, you weigh more. You can find how much more by measuring θ by hanging a weight on a string and then plugging this θ into (6.1).

---

[12] See page 181 for more details.

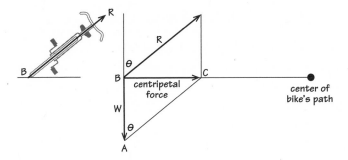

Figure 6.6. The apparent weight gain when going in a circle.

6.9 Feeling the Square in $\frac{mv^2}{2}$ through the Bike Pedals

The kinetic energy K is defined as the work required to bring mass m from rest to a given speed v. It turns out that $K = mv^2/2$, as explained on page 165 in the appendix.

Question. How is the square in $mv^2/2$ felt when pedaling a bike? The biking is taking place on level ground, with no air resistance or rolling resistance.

Answer. Since v is squared, the faster you go, the more "fuel" is needed to gain one additional mile per hour. Indeed, to accelerate from v to $v + 1$ requires the energy

$$\frac{m(v + 1)^2}{2} - \frac{mv^2}{2} = mv + \frac{m}{2} > mv;$$

this energy cost is greater for greater v.

To understand this increasing cost intuitively, imagine pressing on the pedals with constant force, thus accelerating at a constant rate.[13] Thus it takes the same 1 second (say)

---

[13] Friction, air drag, etc., are all ignored. The constant force implies, via Newton's second law, constant acceleration.

to add 1 mph, whether you move slowly or fast. But here is the depressing thing: when your move fast, you have to spin your pedals fast, so in that 1 second your feet will have to travel a longer distance. That means that you will have done greater work in 1 second when moving fast than in 1 second when moving slowly. Summarizing, to maintain constant acceleration, your "engine" has to steadily increase its speed while maintaining the same force. That is, the power output must steadily grow. So, even without friction, life in the fast lane is hard. Friction makes it even harder.

Problem. How much more fuel does it take for a car to reach 70 mph than to reach 10 mph (ignoring all frictional losses and assuming a perfectly efficient engine)?

Solution. *Almost 50 times more!* Indeed,

$$\frac{K_{70}}{K_{10}} = \frac{m70^2/2}{m10^2/2} = 7^2 = 49.$$

6.10 A Paradox with Rockets

An accelerating rocket. When a rocket burns one unit of fuel, it increases its speed by a certain amount. *This amount is the same regardless of the rocket's speed at the start of the burn.*[14] This is because the rocket's acceleration is independent of the rocket's speed. In this way a rocket is different from a bike: the faster a bike goes, the harder it is to accelerate.[15] This difference gives rise to the following puzzling conclusion.

---

[14] All is in weightlessness; speeds are measured relative to an inertial observer.

[15] See page 69 for more explanation.

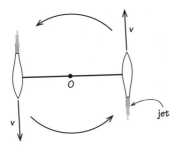

Figure 6.7. The rockets can gain more kinetic energy than the energy stored in the fuel. How is it possible?

A paradox. Figure 6.7 shows two rockets mounted on a rod which can spin freely on the pivot O. We give the rockets an initial spin at speed v and then ignite their engines. At the end of the burn, the rockets' speeds increase by 1 m/s. This speed increase is the same regardless of the starting speed v. The speed now is $v + 1$, and the combined kinetic energy is $m(v + 1)^2$. The kinetic energy gained is

$$\Delta K = \underbrace{m(v + 1)^2}_{\text{after}} - \underbrace{mv^2}_{\text{before}} = 2mv + m.$$

Imagine now that v is a very large speed; according to this formula then the energy gain ΔK can be as large as we like. So the rockets can gain more kinetic energy than what is contained in the fuel! Can this really be right?

Answer. Surprisingly perhaps, the answer to last question is "yes": the rockets can gain more kinetic energy during a burn than the amount produced by the fuel. But this does not violate conservation of energy because something else—the ejected fuel—loses a lot of kinetic energy. When the rocket is fast, this loss is large. If this loss is taken into account, the paradox disappears.

A more detailed discussion of a very similar problem is given on pages 74–76.

71

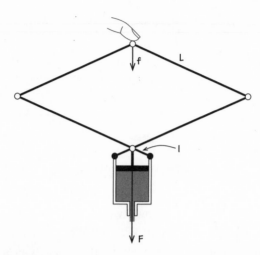

Figure 6.8. Is it possible to cause an object to liftoff the table by pressing it down?

6.11 A Coffee Rocket

Some coffee dispensers have a lever on top. By pressing the lever, you pump coffee down into a cup.

Now the jet of coffee shooting downward creates an upward jet force on the container. This force is of course far too weak to lift the container, not to mention overcoming downward pressure of the hand pressing the lever. This raises the following question.

Question. *In principle,* can the proportions of the dispenser be designed so that it can lift off the table when the lever is pushed down?

Answer (a jumping coffee pot). Strangely, the answer is yes. By making the lever ratio L/l in Figure 6.8 very large, I can achieve a huge compression force F on the piston, and thus a strong jet force, while applying only a weak downward force f (capital letters F, L versus lower case f, l remind us which quantities are larger). So the jet force can exceed the combination of the weight and the downward force f. The catch is that I would have to move the lever's end very fast: in order to maintain high pressure a high enough speed is required (see page 41 on Bernoulli's law). So, unfortunately, the jumping coffee pot is impractical as a practical joke. This is a pity, since, had this idea worked, it would produce an extra surprise: the shooting coffee would push down on the cup with great force—greater than the weight of the pot.

Here is a more detailed explanation of the liftoff idea. For liftoff, the jet force F_J must exceed the weight[16] plus the downward push f:

$$F_J > W + f. \tag{6.2}$$

To be specific, let me push down with the force $f = W/2$; then the above liftoff condition becomes

$$F_J > 1.5W. \tag{6.3}$$

Now it is intuitively clear that if the piston force F is large enough, then the jet force F_J will be large enough to satisfy (6.3). So all we have to do is to produce a very large piston force; this can be done by choosing a large lever ratio. Namely, by the rule of the lever $F = (L/l)f$, and so F can be made as large as we like by making ratio L/l sufficiently large.

Problem. Can the liftoff be achieved if I push down with the force $f > W$, greater than the pot's weight?

---

[16] The weight of the pot is decreasing due to coffee lost, but let us ignore the fine points. For those insisting on more rigor, I can say: let W be the initial, largest weight of the pot, before the coffee started shooting out.

Problem. Water is shooting under pressure from one container down into another container below it, at a steady rate. Both containers rest on the platform of scales. What will the scales register, as compared to the combined weight of the containers and the water?

6.12 Throwing a Ball from a Moving Car

The setting. When sitting in a moving car and throwing a ball forward,[17] I give kinetic energy to the ball. The strange thing is that this energy gain (as viewed by a ground observer) can exceed the energy spent by my muscles. The next paragraph makes all this more precise.

The details. Initially, the ball moves with the car, with speed V. I throw the ball forward with speed $v = 1$; the new speed is then $V + 1$. The change in the ball's kinetic energy is

$$\Delta K = \underbrace{\frac{m(V+1)^2}{2}}_{\text{after}} - \underbrace{\frac{mV^2}{2}}_{\text{before}} = \frac{m}{2} + mV. \qquad (6.4)$$

The paradox. According to (6.4), the faster the car moves, the more kinetic energy the ball gains—for the same throwing speed $v = 1$! Even more surprising, the energy gained by the ball can exceed the energy spent by my muscles if the car is moving fast enough. How to explain this paradox?

Solution. Although (6.4) contains a mistake (see the next paragraph), the strange conclusion is nevertheless correct: the ball can in fact gain more energy than my arm produced. The explanation of the paradox is that this gain comes from

---

[17] Let us assume that the car rolls by inertia, on the horizontal ground, with no friction whatsoever.

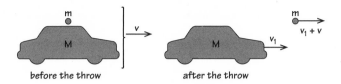

Figure 6.9. Resolving the kinetic energy paradox.

a compensating loss of kinetic energy of the car. We cannot ignore what happens to the car as I did in (6.4), even if this effect may seem small. When I throw the ball forward, the car is pushed backward. So the car's kinetic energy decreases, and when this decrease is counted, the correct total kinetic energy increase is obtained.

Energy balance without cheating. Here is a precise solution of the paradox (assuming no friction, no air resistance, and no other distractions from the main idea). First, we find the speed of the car after the throw. The act of throwing does not change the linear momentum[18] of the system (Figure 6.9):

$$(M + m)V = MV_1 + m(V_1 + v), \qquad (6.5)$$

where M is the car's mass, m is the ball's mass, V_1 is the car's new speed and v is the speed of the ball *relative to the car* when thrown. The total change of kinetic energy[19] is

$$\Delta K_{\text{total}} = \underbrace{\frac{MV_1^2}{2}}_{\text{new car}} + \underbrace{\frac{m(V_1 + v)^2}{2}}_{\text{new ball}} - \underbrace{\frac{(m + M)V^2}{2}}_{\text{old car+ball}}. \qquad (6.6)$$

---

[18] See page 171 for the definition.
[19] Note the difference with (6.4), which refers to the ball only.

Using (6.5), and omitting some algebra, we can reduce this to

$$\Delta K_{total} = \frac{mv^2}{2} \frac{M}{m + M}. \tag{6.7}$$

Discussion.

1. ΔK_{total} does not depend on the car's initial speed V, according to (6.7), just as one would expect.

2. If $M \gg m$, as in the case of the car and ball, then (6.7) gives $\Delta K \approx mv^2/2$ (as if the car were stuck to the ground), again as expected.

3. The change of energy consists of two components:

$$\Delta K_{total} = \Delta K_{ball} + \Delta K_{car} = \frac{mv^2}{2} \frac{M}{m + M}.$$

If the car moves fast, ΔK_{ball} will be large positive, ΔK_{car} will be large negative. That is, the ball gains a lot of energy and the car loses a lot. The net change in kinetic energy does not depend the car's speed, but the way this change is distributed between the car and the ball does.

❀ 7 ❀

PARADOXES WITH THE CORIOLIS FORCE

7.1 What Is the Coriolis Force?

Question. Imagine playing ball on merry-go-round. The merry-go-round is enclosed, so you don't see outside. Standing at the center, you want to hit a target at the rim. You aim the ball straight at the target, but you miss: the ball curves to the right, as in Figure 7.1. Why does this happen? Let us forget gravity. Assume that the spin is counterclockwise.

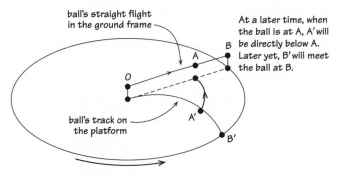

Figure 7.1. Explaining the Coriolis force.

Answer. The shortest answer is: "The ball actually flies straight. But the platform turns, the target moves, and the ball misses to the right of the target. In the enclosed rotating room it appears as if the ball veered to the right."

To describe this in slightly more detail, imagine that the ball marks its path on the platform, perhaps by shooting a jet of ink downward.[1] Although the ball flies straight, the trace it leaves is curved, since the platform spins, as the figure explains. To us on the ground, this curving is no mystery. But the observer on the platform, who feels that the platform is stationary, will have an illusion of an invisible force.[2] This fictitious force is called the *Coriolis force*.

We live in a rotating world, and so the Coriolis force occurs all around us. It causes the rotation of cyclones and anticyclones, and it affects ocean currents. Basics on the Coriolis force can be found in many books, for example Arnold, Goldstein, Landau and Lifshitz.

Question. The Hudson River flows south. In which direction does the Coriolis force push the flowing water?

Answer. The force pushes west. Indeed, imagine a blob of water moving south along a meridian. Because of Earth's spin, everything on Earth moves east, and the farther from the north pole, the faster. And thus as the blob of water in the Hudson moves farther from the pole, its velocity in the eastward direction increases. Resisting this increase due to its inertia, the blob will push against the west bank of the river. Reciprocally, the blob will feel as if it is being pushed by the invisible force toward the east. Does this Coriolis

[1] Again, we ignore gravity and imagine the ball flying in a straight horizontal line.
[2] Earth is such a platform. In fact, for most of human history people were unaware of its rotation.

pressure against the west bank of Hudson explain why the New Jersey shore across from Manhattan is steep, unlike the Manhattan shore? Probably not.

7.2 Feeling Coriolis in a Boeing 747

Question. How strong is the Coriolis force acting upon a person in a jetliner (typical speed: 250 m/sec)?

Answer. To simplify calculation, consider the plane flying near the North Pole. Earth then can be considered a flat disk, a huge merry-go-round, spinning around the polar axis. In this case a passenger feels the Coriolis force[3]

$$F_{coriolis} = 2m\omega v, \tag{7.1}$$

where m is his mass, ω is the angular velocity of Earth, and v is the jetliner's speed. Let's take $m = 70$ kg for a round figure (pardon the pun), $\omega = 2\pi\,\text{rad}/24 \cdot 3600\,\text{sec}$, and $v = 250$ m/sec. Putting it all into (7.1) gives $F_{coriolis} \approx 240$ g, that is, about half a pound!

This force can hold up a cup of water! Another way to get a feel for the magnitude of this force is to see the angle θ by which it would deflect a hanging pendulum. This angle, in radians, is close to the ratio of the Coriolis force to the weight mg:

$$\theta \approx \tan\theta = \frac{2\omega v}{g}.$$

This works out to about 1/600 of a radian, or roughly 0.1°. This is the tilt the airplane must in theory adopt to avoid sideslip. How much difference in height between

---

[3] This formula can be found in the books mentioned earlier. On page 86 I "prove" the same formula without the factor 2, inviting the reader to find the mistake.

the wingtips does this mean? Roughly, the product: (wingspan)·(1/600). With the wingspan of a Boeing 747 of about 60 m, this gives the difference in height of 10 cm. Not a lot—but potentially noticeable.

7.3 Down the Drain with Coriolis

Question. It is often claimed that water in the Northern Hemisphere goes down the drain clockwise, and does so due to the Coriolis force. Is this claim correct?

Answer. The claim is false. There is indeed a Coriolis force, but it is negligible in the toilet or the bath. This force, hard to notice in the jetliner (page 79), is even harder to see in water, which moves thousands of times slower than the airplane. Water spins down the drain for different reasons. Some toilets, for instance, inject water at an angle, so it is already spinning as it enters the bowl. In a draining tub, the spin can occur because the agitated water has some vorticity;[4] this vorticity becomes noticeable only when the water converges at the drain. Another reason why water can start spinning down the drain, even when starting at rest, is a combination of the tub's asymmetry and the viscosity of the water. Figure 7.2 gives an example of the tub which will drain counterclockwise—whether the tub is in Boston or in Buenos Aires.

7.4 High Pressure and Good Weather

Question. High-pressure air masses, called anticyclones,[5] rotate clockwise in the Northern Hemisphere. Why does high pressure go hand-in-hand with clockwise rotation?

---

[4] Vorticity is defined on page 52.
[5] The prefix "anti-" indicates the rotation opposite that of Earth.

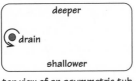

top view of an asymmetric tub

Figure 7.2. Spin down the drain can be caused by an asymmetry of the tub, in combination with viscosity of water.

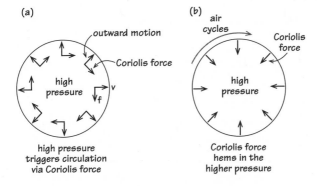

(a)

outward motion

Coriolis force

high pressure

v

f

high pressure
triggers circulation
via Coriolis force

(b) air cycles

Coriolis force

high pressure

Coriolis force
hems in the
higher pressure

Figure 7.3. Higher pressure goes hand-in-hand with anticyclones, due to the Coriolis force.

Answer. The effect is due to the Coriolis force. Imagine air initially spreading from a center, Figure 7.3a. Each particle will feel the Coriolis force trying to deflect the particle to the right.[6] This will cause the particles to veer off the radial path, spiraling clockwise. One can imagine an eventual equilibrium circulation, with the Coriolis force of the circulating particles hemming in the high pressure in the center of the anticyclone, like sheepdogs running around a flock of sheep and compressing it (Figure 7.3).

---

[6] Just like on the blob of water in the Hudson river, page 78.

Why do anticyclones go with fair weather? With the
"pillow" of higher pressure in the center, the air moves
downward, warming up from compression,[7] and the clouds
"melt." The opposite happens in the cyclones, where the air
rises, therefore cooling from expansion, and the moisture
condenses, forming clouds.

7.5 What Causes Trade Winds?

Trade winds form an equatorial band, steadily blowing from
the east. What causes trade winds?

The cause is a combination of convection and the
Coriolis force. The following is a vast simplification, but it
captures the key causes.

1. The colder air sinks southward toward the equator,
   from the higher latitudes; this flow takes place in the
   lower layers of the atmosphere, just like cold winter
   air spills into the room through an open door, along
   the floor.

2. Now this southward air flow is deflected by the
   Coriolis force westward, as Figure 7.4 shows.

3. Once closer to the equator, the air gets heated, rising
   as the result, and moving northward in higher
   altitudes.

The atmosphere is an engine whose fuel is solar energy.
The atmosphere takes up the heat from solar radiation and
then gives heat off to outer space through radiative cooling.
A small proportion of solar heat pumps the motion of the
atmosphere against friction. Friction then is converted to

------

[7] This heating from compression is explained on page 121.

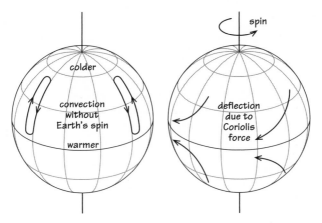

Figure 7.4. Trade winds result from the Coriolis force acting on the convecting air.

heat and is also radiated out. The result is that the solar energy passes through Earth on its way to outer space, but agitating the atmosphere in the process. Earth and all that's on it, including us, is like an organism that digests solar energy and then passes it out in the same amount, although in a different form, or rather in a different part of the spectrum.

❈ 8 ❈

CENTRIFUGAL PARADOXES

8.1 What's Cheaper: Flying West or East?

Problem. A flight east from Boston to London consumes less fuel than the return trip. This is because the jet stream blows roughly toward the east. But what if the jet stream were to magically disappear—would this disparity of fuel consumption disappear as well? To focus on the essentials, let's replace Boston and London by two points A and B on the equator, and ask: In the absence of any winds, would the eastbound trip AB consume the same amount of fuel as the westbound trip BA?

Solution. Going east will take less fuel because of Earth's rotation. Each point on the equator orbits the center of Earth. Going east, the plane goes *with* the rotation of Earth, thus enhancing its orbiting speed around Earth's center. The increased centrifugal force makes the plane a little lighter. And a lighter plane uses less fuel.

How much lighter? For the travel speed of 250 m/sec the weight difference[1] is about 2/3 of 1%. A loaded Boeing 747

---

[1] Worked out in the next problem. The ratio of the weight difference to the actual weight turns out to be $4v\omega/g$, where v is the plane's speed, ω is the angular velocity of Earth, and g is the gravitational acceleration.

can easily weigh 300 tons. This is about 2 tons lighter when going east versus going west—the weight of about 30 people (without luggage)!

In effect, we can think of the plane as a satellite that is just too slow—most of its weight is carried by the wings, and only a small proportion of the weight is supported by the centrifugal force.

On a realistic note, the jet stream has a much greater effect than the centrifugal force.

Problem. What is the proportion of the "centrifugal lightening" of the plane to its weight?

Solution. The difference between the eastbound and the westbound weights is the difference between the centrifugal forces:[2]

$$\Delta W = \frac{m v_{\text{east}}^2}{R} - \frac{m v_{\text{west}}^2}{R} = \frac{m}{R}((\omega R + v)^2 - (\omega R - v)^2);$$

here ω is the angular velocity of Earth, R is the radius of Earth, and v is the plane's speed. Squaring and canceling, we get

$$\Delta W = 4m\omega v,$$

and the ratio to the weight is

$$\frac{\Delta W}{W} = \frac{4m\omega v}{mg} = 4\frac{\omega v}{g}.$$

8.2 A Coriolis Paradox

A person walking with (constant) speed v on a platform rotating with (constant) angular velocity ω experiences the

---

[2] See page 181.

Figure 8.1. What's wrong with this "proof" of (8.2)?

Coriolis acceleration

$$a_{\text{coriolis}} = 2\omega v. \tag{8.1}$$

This is proven in many of the mechanics books mentioned on page 187. Now the next paragraph gives a short "proof" of the same formula, but with "2" missing:

$$a_{\text{coriolis}} = \omega v. \tag{8.2}$$

Question. Can you find where I lost half of the Coriolis force in that "proof"?

"Proof" of (8.2). Imagine me walking along a radius on a spinning platform (Figure 8.1) with speed v. In time Δt I travel from the center to the distance $r = v\Delta t$ from the center. My velocity perpendicular to the radius is now $\omega r = \omega v \Delta t$, due to the rotation of the platform. So my velocity perpendicular to the radius changed by the amount $\Delta v = \omega v \Delta t$ in time Δt; my acceleration is thus

$$\Delta v / \Delta t = \omega v \Delta t / \Delta t = \omega v,$$

proving (8.2). Where is the mistake?

Solution. The velocity shown in Figure 8.1 is wrong; the correct sketch is in Figure 8.2. The "proof" ignored the fact that the radius along which I walked has turned (by $\omega \Delta t$), and with it so did my velocity vector, producing

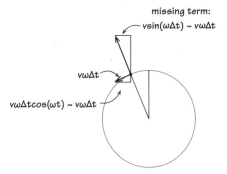

missing term:
vsin(ωΔt) ~ vωΔt

vωΔt

vωΔtcos(ωt) ~ vωΔt

Figure 8.2. Resolving the paradox of the missing half in the Coriolis force.

an additional component $v \sin(\omega \Delta t) \approx \omega v \Delta t$. This is the missing half! In summary, the "2" in (8.1) results from two effects: (1) the difference in the velocities between points of the platform, and (2) the change of the walker's direction due to the turn of the platform.

8.3 An Amazing Inverted Pendulum: What Holds It Up?

Question. The pendulum—a bob on a stick—has two equilibrium positions, of which the top one is unstable: the smallest puff of air will cause the pendulum start falling away. Of course, we can balance the pendulum, as we can a broom on the palm of the hand. Such balancing requires intelligent response to the pendulum's motion. But how will the pendulum behave if we just vibrate the pivot in (say) the vertical direction?

Answer. More than 100 years ago it was discovered that if the pivot point is vibrated fast enough in the vertical direction, then the upside-down position of the pendulum becomes stable.[3] This is a striking fact, and it is completely different from balancing by feedback: the oscillating pivot does not "know" what the pendulum is doing and is not responding in any way to the pendulum's motion. It is quite surprising that vibration should make any difference: after all, why don't the rapid up and down motions just cancel each other? And why does vibration favor stability over instability?

An experiment. Figure 8.3 shows an aluminum rod which can pivot on the blade of a jigsaw (see also a YouTube movie: http://www.youtube.com/user/MarkLevi51#p/a/u/2/ cHTibqThCTU). By starting the jigsaw I cause the blade to vibrate rapidly back and forth in the reciprocating motion—perhaps at 30 cycles per second. It then feels as if an invisible spring were trying to align the rod with the blade. This "spring" is strong enough to keep the pendulum in roughly a horizontal position if I aim the blade appropriately, as illustrated in (b), right.

Question. *Why* does vibration stabilize the pendulum? (No formulas, please.)

Answer. Rather than considering a rod as in Figure 8.3 or in the movie, take the pendulum to be a small bob on the end of a weightless stick. The pivot's acceleration is so violent that I can temporarily ignore the gravity, small by comparison. The bob feels alternately a great pull and a great push from the stick. And since this push–pull force is *exactly aligned with the rod,* the bob wants to move always in that same instantaneous direction of the rod. The bob would thus move in a curved path, as in

---

[3] A. Stephenson, "On a new type of dynamical stability," *Manchester Memoirs* 52 (1908), p. 110.

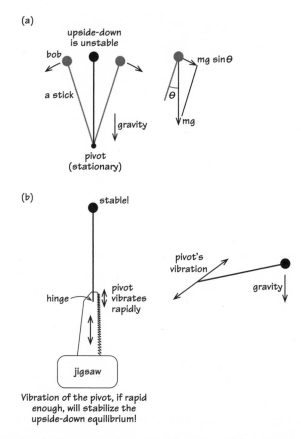

Figure 8.3. (a) The upside-down pendulum is unstable, but (b) it may become stable if the pivot vibrates vertically.

Figure 8.4a. Such a path is called the *pursuit curve,* or the *tractrix.*[4] So let me temporarily constrain the bob to such a tractrix. This is a pretty innocent constraint since I am not interfering with the violent push–pull of the rod.

---

[4] Formally, the tractrix is defined by the property that the tangent segment between it and a given line is of constant length. If I roll the front wheel of a bike in a straight line, its rear wheel will roll on a tractrix.

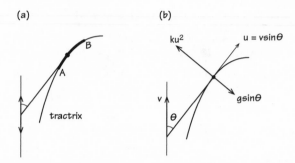

Figure 8.4. A hidden centrifugal force is responsible for the upside-down stability of the vibrated pendulum.

Thus constrained, the bob will execute rapid back-and-forth vibrations along a short arc AB. The arc being curved, the bob will push on it with a centrifugal force, Figure 8.4b. That is, the bob "wants" to go in the direction of that centrifugal force! So if released from the constraint, the bob would obey its desire. If vibrations are violent enough, this force would overwhelm the gravitational destabilizing force[5] and the pendulum would stand upright. This completes our explanation of stability.

The Paul trap. This amazing phenomenon has been known for at least a century; the earliest mention of it known to me is in Stephenson's 1908 paper. The same effect of stabilization by vibration, but in a different guise, is used in the *Paul trap*—a device that suspends charged particles in

---

[5] More details can be found on page 158 in M. Levi, *Physica D* 132 (1999). Remarkably, only a couple of lines are needed to convert the intuitive idea into a stability criterion: the upside-down position is stable if $\langle v^2 \rangle \geq g\ell$, where v is the pivot's speed, and $\langle \cdot \rangle$ denotes the average over the period of the jigsaw; ℓ is the length of the pendulum. Our physical explanation replaces a much longer formal calculation, with the added benefit that it explains "what's going on."

vacuum using vibrating electric fields.[6] Wolfgang Paul was awarded a Nobel Prize for this invention. Paul's explanation of this effect is based on differential equations. This is not an explanation you can give to a random passer-by (and it's better not to try, especially not in a seedy neighborhood.).

8.4 Antigravity Molasses

Question. A jar with a lid is half-filled with molasses or some other honey-like fluid. If the jar is turned upside-down, the molasses will, of course, pour down. But is it possible to make the jar move in such a way that molasses does not pour down even when the jar is inverted (Figure 8.5)? That is, can one balance the molasses upside-down?

Answer. If the jar is vibrated rapidly in the direction of its axis, the molasses will not pour down after the jar is inverted.[7] I reproduced this experiment with simple things

Figure 8.5. Vibration keeps the molasses from pouring out of the inverted jar.

---

[6] In our jigsaw experiment, if I were to place myself in the reference frame of the pivot, I would feel as if the gravity were vibrating.

[7] For a description of an experiment, see M. M. Michaelis and T. Woodward, *American Journal of Physics* **59**(9) (1991), pp. 816–821; a theoretical discussion can be found in G. H. Wolf, *Physical Review Letters* **24** (1970), pp. 444–446.

from the workshop and the kitchen. With the jigsaw turned on, the jar vibrates in the axial direction. As the entire assembly is turned upside-down, the molasses doesn't pour down but miraculously stays inverted, as if the gravity were reversed. Somehow, the vibration keeps the surface of molasses flat, preventing it from pouring down. Even more surprising, if I turn the jar on the side (with the jigsaw still on), the surface of molasses will be vertical, like the wall of the parting Red Sea.

8.5 The "Proof" That the Sling Cannot Work

The paradox. I am twirling a rock on a string. The rock is subject to the tension force T of the string. This force points directly at the point P (pivot), where my fingers hold the rope. Hence the torque[8] of T relative to P is zero. But zero torque means no change in angular momentum. That is, $Lv = $ const., where L is the length of the string and v is the rock's velocity perpendicular to the string. Thus v cannot change—an erroneous conclusion, as Goliath found out. Where is the mistake?

Answer. The implication

$$torque = 0 \quad \Rightarrow \quad angular\ momentum = const$$

is invalid in accelerating frames. And the frame of my fingers is certainly accelerating when I try to twirl the string.

How does the sling work? Figure 8.6 illustrates the answer. In effect, we are creating a pendulum which always slides down the slope, *chasing the bottom equilibrium B which escapes from it.* My hand—the pendulum pivot P—travels

---

[8] These concepts are defined on page 174.

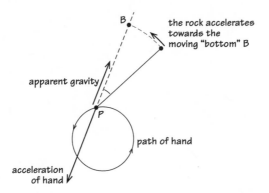

Figure 8.6. The sling is a pendulum chasing the escaping equilibrium.

(say) in a circle, faster and faster. An observer attached to P will feel the g-force as shown. In particular, he will observe the rock accelerating "down" toward the equilibrium B, just like a pendulum accelerates down toward its lowest point. But B "escapes" counterclockwise, and the rock will simply chase B, ever accelerating.

The next paradox takes this question to a surprising limit.

8.6 A David–Goliath Problem

The following paradox arose from the contemplation of the preceding "twirling the rock" problem. The answer to the following question may at first be hard to believe. As mentioned in the previous problem, the sling is essentially a rock on a rope, which is twirled and then released.[9]

The sling problem. I guide one end of the string around a circle in such a way that the rock travels in a bigger concentric circle, with the constant "lead" angle $\theta = 45°$ between the string and the rock's velocity. The rock will

---

[9] The gravity is to be ignored throughout.

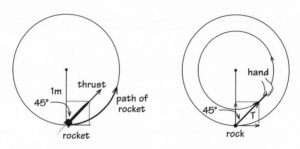

Figure 8.7. Acceleration is in direct proportion to (velocity)^2.

spin ever faster, thanks to the tangential component T_{tan} of string's tension. (I will then have to spin my fingers faster as well, in order to maintain the constant 45° angle.) Assuming the rock travels in the circle of radius 1 m, can you venture a rough guess as to how long will it take the rock to go from the initial speed of 1 m/sec to the speed of sound (330 m/sec)? to the speed of light (300,000,000 m/sec)?[10]

Here is the same problem, restated:

A rocket problem. A toy rocket flies in a circle of radius 1 m, with its thrusters aimed at a fixed angle $\alpha = 45°$ to the trajectory (Figure 8.7).[11] How long will it take the rocket to increase its speed from 1 m/sec to the speed of sound? to the speed of light?

Solution. The rock(et) *will exceed the speed of light—not to mention the speed of sound—in less than 1 second!* In fact, the velocity approaches infinity as the time approaches the 1-second mark. This just means that it is impossible in principle to spin the rock in a circle, forever maintaining the

---

[10] Let's pretend that Newtonian mechanics applies to all speeds, so that things can move faster than the speed of light. Let's also neglect the gravity and the air resistance, as well as the limitations on the strength of the rope and of the human.

[11] This requires the ever-increasing thrust.

constant lead angle of 45° (or any other nonzero angle). Here is an explanation.

An explanation. I will show that the rock's tangential acceleration a is in direct proportion the square of the velocity v—in fact, that $a = v^2$. That is, the rate of change of v is directly proportional to v^2. And by a calculus argument of the next paragraph, such a quantity becomes infinite in finite time.

Precise details. Since the angle between the tension force **T** and the tangent is 45°, we conclude from Figure 8.7 that the tangential and the radial components of **T** are equal. The same is true then for the tangential and the centripetal accelerations: $a = a_c$. But the centripetal acceleration (page 181) is given by $a_c = v^2/r = v^2$ (recall that $r = 1$ m), and we conclude

$$a = v^2. \tag{8.3}$$

And now we need some calculus to show that, as a consequence of this relation, v approaches infinity in finite time. Equation (8.3) amounts to

$$\frac{1}{v^2}\frac{dv}{dt} = 1. \tag{8.4}$$

Taking the antiderivative, we get

$$-\frac{1}{v} = t + c,$$

where we find $c = -1$ by recalling that $v = 1$ for $t = 0$. This gives us

$$v = \frac{1}{1 - t}.$$

95

At the time $t = 0.9999$ sec we get $v = 10,000$ m/sec—enough to launch the rock into Earth's orbit and almost enough to escape the confines of Earth's gravity altogether.

At the time $t = 0.999999$ the rock will exceed the speed of light. At some time before 1 second is up, the rock's kinetic energy will exceed the total energy stored in the Sun, and in all other stars in the Universe. So much for realistic assumptions.

Question. For $t > 1$ we obtain $v = 1/(1 - t) < 0$, which means that the rock is moving backward. How to explain this absurd conclusion?

Answer. For $t > 1$ the formula $v = 1/(1 - t)$ is no longer applicable.

A strange bank account. We saw that if the rate of change dv/dt of a quantity v is directly proportional to v^2, then v becomes infinite in finite time. Imagine for a moment that a bank decides to compound interest on this principle, that is, by letting the balance v change according to such a rule[12]—the interest is directly proportional to the square of the present dollar amount. This would be a dream deal for a customer (and a nightmare for a bank). In particular, the balance would reach infinity in finite time. However, if the customer is foolish enough to wait past that moment ($t = 1$ in our earlier example), the balance will become negative.[13] Suddenly, a huge asset becomes a huge liability (mathematics can imitate life). If something escapes to $+\infty$ in finite time, it can reenter the number line from $-\infty$.

---

[12] That is, by letting $dv/dt = kv^2$, instead of the usual rule $dv/dt = kv$ of continuous compounding.

[13] Provided he uses the solution $v = 1/(1 - t)$ as we derived it, past the blow-up point; the applicability of the formula past the moment $t = 1$ is a matter of debate and would have to be decided by a judge.

With such a strange compounding customers would benefit by pooling their accounts. For example, if two equal accounts merge into one, the interest will quadruple, since $(2v)^2 = 4v^2$, or double the interest per person. This will bring closer the moment when the customers become infinitely rich. Exponential compounding $dv/dt = kv$, the one used in reality, is the only fair type of compounding, meaning that customers neither gain nor lose by pooling their accounts.[14] And one other thing: your profit will be the same whether you measure v in dollars, cents, or euros under the exponential compounding. This is not so for the $dv/dt = v^2$-type: the moment you convince the bank to measure your wealth in cents, you increase your rate of enrichment 100-fold!

8.7 Water in a Pipe

Question. Figure 8.8 shows water flowing through a bent pipe. When approaching the bend, water wants to keep going straight and therefore pushes the pipe in the direction it was going before the turn, as shown by the horizontal force arrow in the figure. Is this horizontal direction correct?

Answer. The figure is wrong: the force actually points "northeast," at $45°$ to both straight sections of the pipe. One should not overlook the fact that since the water turns "south," it must also push on the pipe "north." More formally, let us consider what happens to the linear momentum of a particle of water as it makes the turn. The particle's velocity changes from v to v' (Figure 8.8); the velocity

---

[14] Putting it differently, the differential equation governing the account balance is *linear*. For such an equation the sum of two solutions is also a solution. This means that the pooled account will have the same balance as the sum of the two accounts if kept separate.

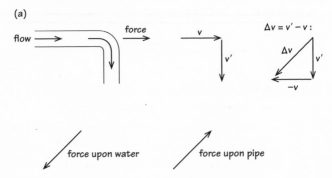

Figure 8.8. In which direction does water push the pipe as it hits the turn?

change $\Delta v = v' - v$ is the bisector of the right angle, as the figure shows. By Newton's second law, the average force acting on this particle is aligned with the velocity change. And by Newton's third law, the water applies equal and opposite force to the pipe.

Another view. Here is an alternative quick way to see that the answer in the figure is wrong. Speaking loosely, the force on the pipe is made up of centrifugal forces of all the particles going through the turn. But the centrifugal force upon a particle is given by mv^2/r, the point being that v is squared. Thus changing v to $-v$ won't change the force. But according to Figure 8.8a the force would be different if the flow were reversed. So the figure is wrong.

8.8 Which Tension Is Greater?

The following problem has a surprising answer, a simple solution, and an even more surprising consequence, described on page 101.

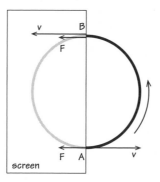

Figure 8.9. Finding the tension of the rope.

Problem. Two closed circular loops of different radii are made out of the same rope, and set spinning with the same speed. Centrifugal effect causes both loops to be under tension. Which loop has greater tension? The rope is assumed to be perfectly flexible and unstretchable, and is not subject to external forces, including gravity.

Answer. The tensions are the same. To see why, and with almost no calculations, let's focus attention on half of the loop only (Figure 8.9) by covering half of the circle with a screen so that we don't have to look at it. We see material injected at A and ejected at B with the same speed v. Between its entrance and exit, each particle changes its velocity by $2v$. The forces causing this change of speed are the tensions F at A and B. To find F, let us wait for some time Δt (it will cancel in the end). During this time Δt a certain mass Δm is injected at A and the same mass is ejected at B. That is, mass Δm changed speed by $2v$ in time Δt. By Newton's second law, $F = ma = m \Delta v / \Delta t$, we have

$$(2F)\Delta t = \Delta m \cdot (2v),$$

99

or

$$F = \frac{\Delta m}{\Delta t} v.$$

Now $\Delta m = \rho \cdot (v \Delta t)$, where ρ is the linear density (mass per unit length). Substituting this into the last expression gives

$$F = \rho v^2,$$

so indeed the tension does not depend on the radius of the circle—only on v and ρ.

An even shorter proof of the independence of the tension F on the radius focuses on a semicircle, noting that the force $2F$ keeps its center of mass in a circular orbit:

$$2F = \frac{mu^2}{r},$$

where m is the mass of the semicircle and r is the distance of its center of mass to the geometrical center. It then remains to note that (1) m/r does not depend on the radius R since both m and r are directly proportional to R, and (2) u does not depend on R (only on v). Therefore, F does not depend on R either.

8.9 Slithering Ropes in Weightlessness

Ropes can behave in very surprising ways. Here is one. If a rope[15] is formed into a circular loop and set spinning (Figure 8.9) in weightlessness, it will keep its circular shape.

Question. Do any shapes besides the circle remain unchanged, with the particles circulating along the curve? For

---

[15] Our rope is idealized: it does not stretch, does not resist bending, and is very thin. A good example is the chain made of little balls, of the kind that tether pens at banks.

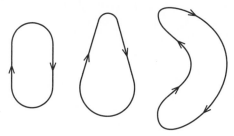

Figure 8.10. The initial velocity of each particle is tangent to the curve. Which if any of these shapes will remain unchanged in weightlessness when set in motion as shown by arrows?

instance, will some of the shapes in Figure 8.10 remain unchanged when given initial velocity tangent to the curve? Equivalently, imagine a perfectly flexible water-filled hose, with water set circulating around the hose without friction. Which shapes in Figure 8.10 remain fixed?

Answer. Hard as it may be to believe at first, *any* smooth shape will remain unchanged under the assumptions of the last paragraph.[16]

An explanation. The reason for this strange fact is revealed once we recall (from page 100) that the tension of a *circular* rope is independent of the radius. But we can approximate any curve by concatenating circular arcs of different radii. And since the tension in each of these arcs will be the same, the rope will be happy to keep its shape.

A more rigorous explanation. This requires some calculus. Wishing to apply Newton's second law to the moving rope, let us label each particle of the rope by its distance s

---

[16] E. J. Routh, *Dynamics of a System of Rigid Bodies*, Part 2, 4th ed. (London: MacMillan and Co., 1884), pp. 299–300.

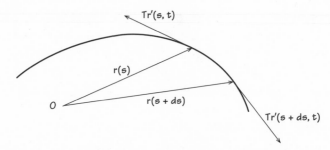

Figure 8.11. Newton's second law for the moving chain.

along the rope from a marked point. Let's denote by $\mathbf{r}(s, t)$ the position vector of this particle at time t (Figure 8.11).

As I show in the next paragraph, Newton's second law states

$$\rho\ddot{\mathbf{r}} = (T\mathbf{r}')', \tag{8.5}$$

where each dot indicates a time derivative, $' = \partial/\partial s$, and $T = T(s, t)$ is the tension of the rope. In addition, the rope is inextensible:

$$\|\mathbf{r}'\| = 1, \tag{8.6}$$

where $\|\cdot\|$ denotes the length of a vector. Equations (8.5)–(8.6) form a complete system for the unknown functions $\mathbf{r}$ and T. Now it is easy to show that any "slithering" motion of the rope retains shape as claimed before. Take any closed curve and parametrize it by the arc length: $\mathbf{R} = \mathbf{R}(s)$. If a particle starts at $\mathbf{R}(s)$ and slides along the curve with speed v, then at time t it will be at $\mathbf{R}(s + vt)$. If we substitute $\mathbf{r}(s, t) = \mathbf{R}(s + vt)$ and $T = \rho v^2$ into (8.5)–(8.6), we get an identity, as seen by direct verification. This proves the claim.

It remains to derive (8.5). Exactly two forces are acting on the arc $(s, s + ds)$—the tension forces on its ends. Their resultant is $(T\mathbf{r}')_{s+ds} - (T\mathbf{r}')_s$, where the t-dependence is suppressed in the notation. The center of mass of the arc is the average position: $(1/ds)\int_s^{s+ds} \mathbf{r}(\sigma, t)d\sigma$; the acceleration of the center of mass is the second time derivative: $\mathbf{a} = (1/ds)\int_s^{s+ds} \ddot{\mathbf{r}}(\sigma, t)d\sigma$. The mass of the arc is $m = \rho ds$; here ρ is the linear density, that is, mass per unit length. Newton's law $ma = F$ becomes

$$(\rho ds)\frac{1}{ds}\int_s^{s+ds} \rho\ddot{\mathbf{r}}(\sigma, t)d\sigma = (T\mathbf{r}')_{s+ds} - (T\mathbf{r}')_s.$$

Dividing both sides by ds and taking $ds \rightarrow 0$ results in (8.5).

Here are some interesting facts/problems for the reader familiar with vector calculus:

1. Show that the angular momentum of the "slithering" motion of the planar rope equals $\rho v A$, where v is the speed and A is the area enclosed by the rope.

2. Show that the z-component of the angular momentum of the "slithering" motion of a spatial rope is $\rho v A_{xy}$ where A_{xy} is the (signed) area of the projection of the rope onto the xy-plane. A similar statement then applies to any straight line and a plane perpendicular to it.

3. Define the *circulation* of the rope as the integral $\int v_{\text{tangent}}\, ds$, the integral of the tangential speed against the length of the curve. Show that for *any* free motion of the rope (not only the "slithering" motion) the circulation remains constant, assuming the rope is free in weightlessness.

103

GYROSCOPIC PARADOXES

9.1 How Does the Spinning Top Defy Gravity?

What holds the spinning top upright is not some force that
acts *against* gravity. Rather, it is a strange *deflecting* force—
the force that stays always *perpendicular* to the direction
of the motion of the axis of the top. This diverting force
"subverts" the instability: the top begins to fall, but then
veers off, thus moving as shown in Figure 9.3. In the next
paragraph I will give a feel of how this strange "gyroscopic
force" comes about from Newton's second law.

A gravity-defying bike wheel. Let's use the bike wheel as
our spinning top. Hang the bike wheel on two strings as
shown in Figure 9.1, and spin it rapidly. Now, cut one string.
Amazingly, the unsupported end of the axle will not drop
down.[1] Instead, it will start a slow precession. The faster the
wheel's spin, the slower the precession will be.

Question. What makes the wheel stay up, defying gravity?

---

[1] Provided you spun the wheel fast enough.

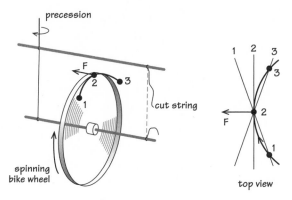

Figure 9.1. The gist of a gyroscope: inertia of some particle holds the wheel up after the string is cut.

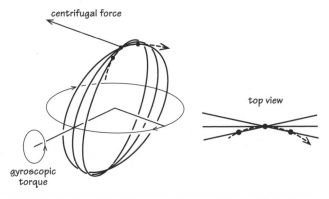

Figure 9.2. The anatomy of the gyroscopic effect—a close-up.

Answer. In a nutshell, a certain *centrifugal force* is responsible—not one that may first come to mind, but rather an orthogonal one!

Let us watch the spinning wheel as its axis precesses. Figure 9.2 shows the path of a typical particle as it passes

105

near the top of the wheel. This path is curved due to the axle's reorientation. The particle, which wants to go by inertia as straight as possible, will resist the deflection with some centrifugal force F, as shown. A similar force $-F$ is exerted by a particle near the bottom. The combined effect is as if an invisible torque were twisting the wheel around the line L; it is this torque that keeps the wheel from dropping.

A strange force. The spinning wheel gives an example of a very strange force akin to magnetic force on a moving charge: unlike friction, this force is *perpendicular* to the motion of the axis. To be more precise, let's turn our wheel into a spinning top by placing one end of the axle on the ground (so it cannot slide but can pivot), as in Figure 9.3. Let us see how would it feel if we were trying to move the free end A of the axle; to avoid confusing the issue, we ignore gravity.

Problem. In which direction must I push the end A of the axle of a spinning top in order to move A with constant speed?

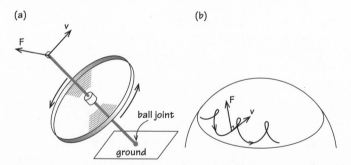

Figure 9.3. (a) A steady motion of the axle requires a steady force in the perpendicular direction. (b) The constant deflecting action of the gyroscopic effect prevents the top from falling.

106

Solution. I must apply the force perpendicular to the desired direction of motion (Figure 9.3)! The explanation has been given in the discussion of the spinning and precessing bike wheel. Playing with an actual spinning wheel gives a strange feeling: as you push on the axle, it responds by moving at the right angle to your push. Once you recognize this behavior, it becomes easy to reorient the wheel in any direction you like, without much effort.

The magnetic force upon a moving charge acts in the same way, perpendicular to the velocity of the charge.

On a lighter note, the gyroscopic/magnetic effect has an analog in human psychology: some people react to a stimulus in the direction orthogonal to the applied stimulus, although this is not what is referred to as a magnetic personality.

Stability by deflection. The spinning top stays up not by resisting gravity, but more subtly. Any motion of the axle generates the gyroscopic force[2] perpendicular to the motion, as illustrated in Figure 9.3. As the figure shows, the top may be falling initially, but it gets deflected from a downward fall. The constant action of this deflecting force results in a path of the kind illustrated in Figure 9.3. One could call this mechanism "stability by deflection."

Energy considerations. The fact that the axle reacts with a force *perpendicular* to imposed motion (Figure 9.3) can be explained by conservation of energy, as follows. If I move the end of the axle with constant speed v, I do not change the gyroscope's energy of spin. Indeed, the bearings are perfect, so I do not speed or slow the gyro's rotation. Therefore,

---

[2] To be clear, this force is fictitious. When I say "force" I mean that the top behaves as if an external force were acting upon it.

I do no work, and hence the force of my hand must be perpendicular to the velocity of my hand.

9.2 Gyroscopes in Bikes

A modern bike is a product of a long Darwinian technological evolution, and is probably as close to perfection as anything civilization created. It is much easier to learn to ride the bike than to explain via Newton's laws how we do it.[3] The following two problems address this very complex question.

Question. While riding the bike straight, I start gently turning the handlebars to the right. How does the gyroscopic effect of the front wheel manifest itself?

Answer. The effect will be to tilt the frame to the left; this is explained by Figure 9.4a.

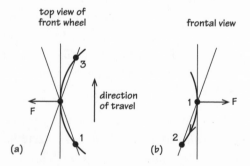

Figure 9.4. Gyroscopic effect of the front wheel. (a) Turning the wheel to the right tilts the bike to the left. (b) Tilting the bike to the left turns the wheel to the left as well.

---

[3] This shows, some would say, that the body is smarter than the mind.

Question. When riding the bike hands-free in a straight line, I tilt the frame to the left—say, by bending my body at the waist. Which way will the gyroscopic effect try to turn the front wheel?

Answer. Left as well, as explained by Figure 9.4b.

9.3 A Rolling Coin

The rolling coin's ability to avoid falling seems magical. It feels as if the coin has the intelligence, or at least the reflexes, of a monocycle rider—in fact, it is better than an inexperienced one! All the random bumps on the surface do not matter—the coin adjusts to them; even if you nudge it gently, the coin will react and stay upright somehow. With no moving parts, the coin is the ultimate simplicity; what explains its "reflexes"? How does the brainless metal disk make the infinitely many "decisions" and adjustments, navigating the tightrope between falling left or right?

The following problem takes the first step in explaining this magic. But even after understanding how it all works, I cannot shake off the feeling of wonder at the natural design that makes a rolling coin stay upright. Despite the explanation (given soon), it still feels like a lucky accident that the gyroscopic effect *helps* the coin stay up rather than, on the contrary, pushing it down. The contrast between the coin's ability to stay upright and its intelligence seems striking.[4]

Problem. As a coin is launched forward with a slant to the right, as in Figure 9.5, its path curves to the right as well. This curving to the right is a "lucky coincidence": without it,

---

[4] This could be said to apply to some individuals as well.

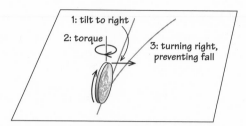

Figure 9.5. The rolling coin adjusts its path to prevent the fall, thanks to the gyroscopic effect. Gravity due to tilt (1) causes the gyroscopic torque (2) and results in coin turning right (3).

the coin would have fallen. As if endowed with a brain, the coin that's tilted right turns right—just like a leaning biker does—thus preventing a fall. Why does the coin turn in the direction of its lean?

Solution. As a coin is launched forward with a right tilt, it starts falling (Figure 9.5). This falling—that is, the coin's growing tilt—results in the gyroscopic torque (explained in the next sentence; see also page 105) which turns the coin to the right, causing the path to curve to the right; this "corrective measure" prevents the fall. To explain this gyroscopic torque, imagine what would have happened if the coin were traveling straight while just tilting more and more. From the reference frame traveling alongside the coin, we would have observed the curving of paths of particles on the rim as shown in Figure 9.2. The resulting centrifugal force will try to turn the coin to the right.

This explanation only suggests that the coin will not fall, but does not really prove it. How do we know, for instance, that this self-correcting effect is strong enough to keep the coin upright? Or that this effect does not overcompensate, causing some kind of unstable oscillation? In fact, the coin has to roll fast enough to stay upright. A discussion of stability of a rolling coin can be found on pp. 55–63 of *Dynamics of Nonholonomic Systems* by Neimark and Fufaev.

9.4 Staying on a Slippery Dome

Problem. Can a solid ring be placed on a perfectly slippery hemispherical dome (Figure 9.6) so as not to slide off, even if nudged? Balancing the ring precisely on top of the sphere does not count, since a small nudge will cause the ring to slide off. No external supports are allowed, including magnetic or other devices.

Solution. Place the ring on the dome, spin it fast around the central axis, and release it. If spun fast enough, and if placed not too far from the top, the ring will not slide off,[5] moving as in Figure 9.7d.

Why it works. Our ring on the sphere is essentially a spinning top: a wheel with a long axle pivoting on a ball joint (Figure 9.3a, page 106). To explain this analogy: instead of

Figure 9.6. Staying up on a perfectly slippery dome.

---

[5] The friction is absent; otherwise the ring would slow down and eventually slide off.

building a slippery sphere (very impractical) we can simply attach the ring to one end of a pin by weightless spokes— essentially a bike wheel with a long axle. The end of the axle rests on the table or is attached to the table by a frictionless ball joint. The ring will thus be effectively confined to an invisible sphere, with no friction. So the ring on the sphere is just a spinning top.[6] And a top doesn't fall if spun fast enough. The reader can refer to the explanation on page 105, or read the next paragraph.

A direct explanation. Shortly after its release, the ring starts sliding down. Hence, viewed from the reference frame of the ring's center, the ring appears to turn as shown in Figure 9.7b. This turning causes the curving of paths of particles of the ring—for example, the path 1-2-3. Due to inertia, the particle resists this curving with a centrifugal force F perpendicular to the plane of the ring. The antipodal particle of the ring exerts an equal and opposite force F'. These two forces exert the torque upon the ring, Figure 9.7c. (I considered only two particles, but others, combined, have a similar effect.) This gyroscopic torque makes the ring veer from the direction it's moving in. As the result, the ring will move as shown in Figure 9.7d.

This is how the fall is prevented: not by direct resistance, but rather by diversion!

For an interesting exercise of physical intuition, imagine your motion if sprawled on a slippery dome and given a considerable cartwheeling spin.

---

[6] For the spinning top analogy to be complete we must also assume that the ring cannot lift off, *which it in fact may do*. So let us assume that the ring is constrained to the sphere, perhaps by some magnetic arrangement.

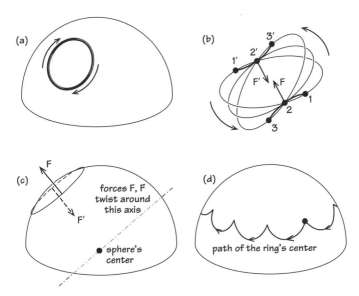

Figure 9.7. How does the ring manage to stay on the slippery dome.

9.5 Finding North with a Gyroscope

Question. How can you find the direction due north using a gyroscope? The gyroscope is frictionless and can spin forever.

Answer. Mount the gyroscope horizontally on a platform, and float the platform in a tub of water. The gyroscope's axis will slowly orient itself exactly along the meridian! And the north direction is the one which the gyro tries to corkscrew into, according to the right-hand rule: if the axis were a screw, it would turn so as to advance north. Putting it differently, the gyroscope tries to align itself with Earth's rotation as much as possible subject to the gyro's axis being constrained to the horizontal plane.

113

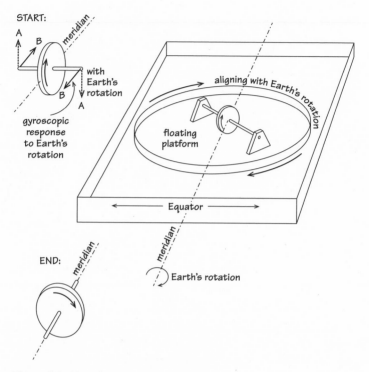

Figure 9.8. How the gyrocompass works.

Why does the gyro seek north? For simplicity, let us place the mechanism on the equator (Figure 9.8). Due to the rotation of Earth, the entire assembly in effect rotates around the north–south dotted line. Let us imagine the gyroscope's axis to be initially in some other direction—say, East–West, as in the figure. Due to Earth's rotation the axis is pushed along the arrows (A); the gyroscope responds by reorienting in the direction (B). In this, the gyroscope behaves just as explained on page 106: when the axis is pushed in one direction, it responds by moving in the perpendicular direction. Eventually, the axis of the floating gyro orients itself along the meridian, as the figure shows.

More freedom. Instead of floating the gyroscope on a platform, one can leave the axis unconstrained by mounting the gyro on gimbals or immersing it in fluid so as to make it neutrally buoyant. The gyroscope will then gradually align itself with Earth's axis. And the angle of the axis with the horizontal immediately tells us our latitude!

In summary, whatever the method of suspension, the gyroscope tries to conform its rotation to that of Earth, as much as the constraints allow.

Problem. Why does the suspended gyrocompass align its axis with that of Earth? (Hint: the friction with the fluid causes the gyroscope to reorient itself.)

The beauty of the gyrocompass is its independence of magnetic anomalies—a big advantage on a steel ship—as well as its ability to find geographic, rather than magnetic, north.

Some history.[7] The gyrocompass was patented in 1908 by E. A. Sperry, the author of numerous other inventions, with over 400 patents in his name. The gyrocompass is perhaps the best known of Sperry's inventions and it played a major role in World War I. After Sperry's death, a United States Navy ship (a submarine tender) was named after him. The USS *Sperry* was launched just ten days after the Pearl Harbor attack. After a long service, which ended in 1982, the *Sperry* became a museum ship. Sperry's gyrocompass is used on ships to the present day.

---

[7] See http://en.wikipedia.org/wiki/Gyrocompass for more details and references.

Sperry, together with Peter Hewitt, also invented an unmanned aerial vehicle—an airplane drone—back in 1916, in the middle of World War I. Sperry prophetically referred to the drone as "the bomb of the future." For the rest of the war, in a long series of trials and errors, Sperry and Hewitt attempted to make the concept work more reliably. During those trials, Lawrence Sperry, the inventor's son, risked his life in some extremely scary test flights, with some close calls.

※ 10 ※

SOME HOT STUFF AND COOL THINGS

10.1 Can Heat Pass from a Colder to a Hotter Object?

The answer to the question is, of course, no: in contact between two objects, the heat goes from the hotter object to the colder one.[1] So the following question may sound silly:

Problem. Can a glass of $100°$ C water heat a glass of $0°$ C milk to more than $50°$ C, their common temperature if mixed together? The glasses are of the same size. Let's also assume that water and milk are completely identical in all their thermal properties.[2] No heat can be brought in from the outside, *but additional containers can be used.*

Solution. Such heat transfer is possible without violation of the second law. All we need is an extra empty glass and a small ladle. Scoop up a ladle of cold milk, dip it into hot water, and wait for the temperatures to become practically

---

[1] This is a consequence of the second law of thermodynamics, an experimentally established fact.

[2] In particular, they have equal specific heat capacities: equal amounts of heat give equal masses equal temperature increases.

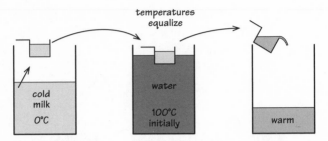

Figure 10.1. There are N scoops in a full glass. Each successive scoop of $0°$ milk decreases the water temperature by the same factor of $1/(1 + 1/N)$.

the same. Pour the ladle into the empty glass. Repeat the process until all the milk is in the third glass. On its way to the third glass, each ladleful of milk picks up some heat from the water. I claim that once thus transferred, the milk will be warmer than the water. To see why, imagine the picture just as you lifted the *last ladleful* out of the water glass. This ladleful of milk is at the same temperature as the water. And the rest of the milk is warmer than that since the earlier ladlefuls are warmer. This shows that once we dump the last ladle in the right glass, the milk will be warmer than the water.

Cooked salmon and Euler's number e. We saw that this method gets milk to over $50°$ C, but how much over? It turns out that with small enough ladles one gets to about $63°$ C, too hot to drink—in fact, a perfect internal temperature of cooked salmon.

What is particularly interesting from the mathematical point of view is that the theoretical best temperature via the ladle method is

$$\frac{100°}{e}, \qquad e = 2.718\ldots,$$

where e is Euler's number from calculus, the limit of $(1 + 1/N)^N$ as $N \to \infty$. Here is why. Let us assume that the glass holds N ladles, where N is any integer we like. A ladleful of cold $0°$ milk dipped in the water glass of temperature T will settle to a shared temperature $T/(1 + 1/N)$ – this is because with the addition of one cold scoop to N warm ones the entire heat of N warm scoops is "diluted" among $N + 1$ scoops, and so the heat per scoop decreases by the factor $N/(N + 1) = 1/(1 + 1/N)$. The temperature then decreases by the same factor. This temperature decrease happens from each step to the next:

$$T_{k+1} = \frac{T_k}{1 + 1/N}, \quad T_0 = 100°.$$

Since after N steps the initial water temperature will have been repeatedly divided by the same quantity N times, we obtain

$$T_N = \frac{100°}{(1 + 1/N)^N} \approx \frac{100°}{e}.$$

Body temperature. For large N one has $(1+1/N)^N \approx e = 2.718\ldots$, and so $T_N \approx 100°/e \approx 36.8°$. By a remarkable coincidence, this is close to the normal temperature of a human body. If you have to calculate e in an emergency and happen to have a thermometer, you can measure your temperature in Celsius and substitute it in

$$\frac{100°}{T_{\text{human}}} \approx e$$

(the estimate will be on the low side if you run a fever, or on the high side if you have hypothermia). This observation makes the natural logarithm—the one with base e—seem even more natural.

119

Once we are on the subject of coincidences, here is another one, relating the temperature of human body and that of perfectly cooked salmon (63°):

$$T_{\text{human}} + T_{\text{salmon}} \approx 100°.$$

Doing better than $63°$ **C.** It turns out that one can do better: a near-perfect swap between the temperatures of the two liquids can be executed, at least in theory. To do so we have to break *both* liquids, not just the milk, into small portions. More practically, we can pass water and milk in opposite directions through two tubes in close thermal contact, as in Figure 10.2. If we pump the milk from the left and the water from the right as shown in Figure 10.2, the temperatures come close to a perfect exchange. This simple device is called the (countercurrent) heat exchanger.

Humans must have borrowed the idea of heat exchanger from nature. In fact, we have heat exchangers in our arms, where deep veins run alongside the arteries. In cold conditions the cold blood from the hands returns through these veins, picking up the heat from the outbound arterial blood. The warmed incoming blood helps keep up the body's core temperature. And the now cold arterial blood going to the limbs does not give off as much heat to the outside. In hot conditions this mechanism is turned off: the inbound blood

Figure 10.2. The countercurrent heat exchanger allows a near-perfect temperature swap.

takes a different route along the surface veins, helping to dissipate the unwanted heat.

Heat exchangers help dogs, sheep, camels, and other animals keep their brains at cooler temperature than the rest of the body: the cooler venous blood from the mouth and nose cools the arterial blood supplying the brain (the organ most vulnerable to overheating). Rabbits, which have no such mechanism, are at a disadvantage in this respect and can die of overheating if chased by a dog in hot weather for too long. Gray whales have numerous countercurrent heat exchangers throughout their tongues (the tongue cannot be insulated by blubber). And this mechanism is not limited to mammals: some fish, for example, tuna, use countercurrent heat exchangers to keep muscles as much as 14° C above the water temperature.

10.2 A Bike Pump and Molecular Ping-Pong

Question. A bike pump gets hot when used to inflate a tire. Is this heating due to friction or to something else?

Answer. Friction is not the main reason. Rather, it is the compression that heats the air inside the pump; this air heats the pump's walls.

Question. Why does compression heat up the air?

Answer. Ping-pong and tennis players all have held the answer in their hands, as it were. The tennis racket hitting an incoming ball is like a moving piston of the pump hitting an oncoming molecule. Thanks to the racket's forward motion, the ball gains speed after the collision (Figure 10.3). (In fact, the speed gained equals twice the speed of the oncoming racket, provided the collision is perfectly elastic and the

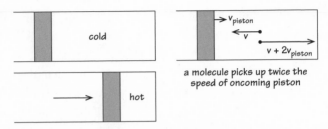

Figure 10.3. A molecule increases its speed by twice the speed of the oncoming piston.

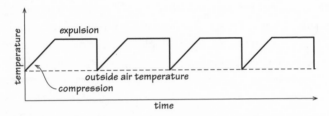

Figure 10.4. Air temperature inside a bike pump as a function of time.

ball's mass is small compared to the racket's.) In the same way, the molecules meeting the advancing piston pick up speed; in other words, the air becomes hotter.

Figure 10.4 illustrates how the temperature of the air inside the pump changes with time. The graph shows that the average temperature is higher than the ambient temperature. The pump's wall settles to some kind of average temperature.

10.3 A Bike Pump as a Heat Pump

The heat pump is simply the refrigerator used as a heater. The refrigerator "pumps" the heat from its inside to the room outside. People who refer to the refrigerator as a heat pump simply express their interest in the heating function of the device rather than in its cooling function—but the two functions are just two inseparable sides of one coin.

Question. How can a bike pump be used as a heat pump to extract heat from cold outside air and give it to warm indoor air? I am not concerned with practicality here, but just with the principle.

Answer. A bike pump can be used to "pump" the heat from the cold outside air into the warm house air, as follows. A pump with the outlet plugged up is just a piston in a cylinder, filled with air (Figure 10.3). Initially the pump is outdoors, uncompressed, and cold. I then depress the piston enough to cause the compressed air to become warmer than the indoor temperature (compression causes heating up), bring the pump indoors and let it give off some heat to the indoors. Once it is about the room temperature, I rush outside and release the piston; the expanding air becomes colder than the outside air. This is so because it gave off some of its heat to the indoors. Being extra cold, the pump now sucks up the heat from cold winter air! This heat replaces the heat we gave to the indoor air. Once the pump reaches the outside temperature, the cycle is complete and can be repeated indefinitely. In actual heat pumps the principle is the same, although the implementation is of course much more intelligent than my running back and forth. A refrigerant is used instead of air; compression/expansion is replaced by condensation/evaporation of the refrigerant. The refrigerant is pumped through pipes, rather than being carried. But the principle is exactly the same.

Efficiency. Surprisingly, the heat pump uses less energy than burning fuel, for the same amount of heating. This is so because part of the work I did in compressing the piston when outdoors is returned to me later when I return the pump to the outdoors and release the piston; the piston will thus give me back some of the energy I had spent compressing it.

10.4 Heating a Room in Winter

Question. One of the two otherwise identical rooms in a house is warmer than the other. Do the air molecules in the warmer room have greater combined kinetic energy than the air molecules in the colder room?

Answer. The energy is the same![3] Although individual molecules have more energy in the hotter air, there are fewer of them since the heated air expands and some escapes through the cracks. As it turns out, these two opposing effects (faster but fewer) cancel out. The cancellation happens because the number of molecules in the room is inversely proportional to the temperature T (counted from absolute zero), while the kinetic energy per molecule is directly proportional to T. In the next paragraph I restate all this more formally.

Explanation, more formally. According to the ideal gas law (a reasonably good approximation at the temperatures in question),

$$pV = NkT, \qquad (10.1)$$

where p is the pressure in the room, V is the room's volume, and T is the absolute temperature[4] of the air, N is the number of molecules in the room and k is a constant independent of

---

[3] I ignore secondary effects such as the expansion of the walls due to heating and the impossibility to keep the temperature exactly constant throughout the room.

[4] That is, the temperature counted from absolute zero.

the quantities mentioned above (k is called the Boltzmann constant).

On the other hand, the kinetic energy E of an average molecule is known to be directly proportional to the temperature of the gas: $E = (3k/2)\,T$. The total kinetic energy of all N molecules in the room is then

$$E_{\text{total}} = NE = N\frac{3k}{2}T = \frac{3}{2}NkT \overset{(10.1)}{=} \frac{3}{2}pV.$$

Since both p and V remain constant as we heat the room, so does E_{total}, as claimed.

10.5 Freezing Things with a Bike Tire

Question. How can a freezing temperature be created on a hot summer day using a bike tire?

Answer. All you have to do is to press the release valve on an inflated tire. Say your pressure gauge shows 3 atmospheres. This really means above the ambient pressure, so in reality it's 4 atmospheres. The escaping air goes from 4 to 1 atmospheres and so it expands a lot; the expansion causes cooling. With this drop in pressure the absolute temperature[5] of the air drops by the factor 2.5 or so. A warm 80° F, or 300° K, cools to 120° K $\approx -244°$ F! This neglects heating due to viscosity when passing through a narrow opening, but one can bet that the escaping air will still be at least below freezing temperature.

The calculation. Consider a small blob of air that moves, in a very short time, from inside to outside the tire. The whole thing happens so fast that we can reasonably neglect heat exchange between the blob and the neighboring air. For

---

[5] That is, the temperature relative to absolute zero.

such expansion without heat exchange (called the adiabatic expansion), the temperature of the air is in direct proportion to the 2/5 power of the pressure:

$$\frac{T_2}{T_1} = \left(\frac{p_2}{p_1}\right)^{\frac{2}{5}},$$

where the subscripts indicate the initial and final temperature and pressure. Now $p_2/p_1 = 1/4$; raised to power 2/5 this gives about 0.57. With $T_1 = 300°$ K we get approximately

$$T_2 = 300° \cdot 0.57 = 171° \text{ K},$$

or about $-152°$ F! By pressing an air release valve you create the coldest place on Earth (outside a cryogenic lab), although a very tiny one and for a short time.

⚘ 11 ⚘

TWO PERPETUAL MOTION MACHINES

The perpetual motion machine is a utopian dream. Like other utopias, it attracts its share of cranks. Fortunately, in contrast to many social utopians of the past (and, unfortunately, present), these tend not to be dangerous and do not usually kill for an idea. Common to all utopias is an attempt to break a law—be it the law of conservation of energy, a law of economics, a law of human psychology, or a law of society.

The inventor of a perpetual motion engine must have limitless intelligence to rise to an infinitely difficult task of inventing the impossible. Attempted inventors of a perpetual motion machine include some very smart people, but few wise ones.

The two machines proposed here are puzzles, each asking to uncover a hidden flaw.[1]

---

[1] In physics, debunking false theories became a safe activity soon after Galileo. In economics things took longer in some countries. A person I knew was sentenced to 12 years of hard labor in the former Soviet Union (around 1947) for wondering aloud in an economics class whether the profit motive is a necessary ingredient for a well-functioning economy.

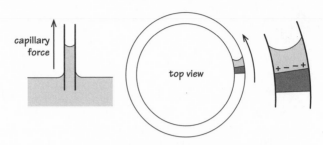

Figure 11.1. A capillary perpetual motion machine.

11.1 Perpetual Motion by Capillarity

Water rises in a thin straw by capillary action, as Figure 11.1 illustrates. Can we harness the work done by rising water? The rising water could pull something, thus doing work, but unfortunately the "engine" stops once water reaches a certain height. Here is a proposed way around the problem. Let's place a piston inside a thin endless tube and put a drop of water against the piston as shown in Figure 11.1, with no bubbles between the water and the piston. The piston can slide with negligible friction. Let's place the tube on the horizontal table to avoid fighting gravity. The same capillary effect as in the straw will pull the piston along the tube, but now there is nothing to stop the piston. The piston will then circulate around the tube indefinitely. For this system to work, all I need is to make sure that the friction of the piston is smaller than the capillary force pulling it along. We secured an infinite source of energy (the small but nonzero friction generates some heat), requiring no fuel.

Question. Where is the mistake? Unless this is the first working perpetual motion machine, something must be wrong with this argument. What is it? That is, where is the mistake in my "proof" that the piston will circulate? (It is not in the assumption of small friction.)

Answer. The problem with our engine is more serious than poor lubrication. Let us first review how water climbs up the straw. Capillarity works thanks to two effects: (1) surface tension – the water surface behaves as a stretched rubber membrane due to an arrangement of water molecules; and (2) the electrostatic attraction of the water to the walls of the tube. The electrostatic attraction pulls the water toward the tube's inner wall, causing the surface to acquire the concave shape of the meniscus. Immediately, the surface tension comes into play: trying to flatten the concave surface, it drags the water column behind it. These two processes— spreading along the wall and dragging—happen simultaneously, and the water climbs up the straw. Eventually the gravity becomes too much and the climb stops.

Now returning to our engine, the same effect that pulled the water up the straw is still present: the concave meniscus still pulls the water, trying to drag the piston. However, we did not consider what happens near the piston—and there is, it turns out, an opposing effect that kills our idea. Due to the electrostatic attraction, the water pressure near the walls is higher, and, in particular, it is higher near the piston. This extra pressure pushes the piston against the desired direction. This is an opposing effect which I initially overlooked.

11.2 An Elliptical Mirror Perpetuum Mobile

This puzzle, which I learned from Peter Ungar in the mid-1970s, relies on the beautiful property of elliptical mirrors: any ray of light emitted from one focus of an ellipse will,

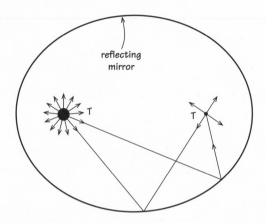

Figure 11.2. Each sphere radiates light. The inside of the ellipsoid is perfectly reflective.

after reflection from the ellipse,[2] pass through the other focus. The following idea of a perpetual motion machine uses this property.

A hot idea. The inside of an ellipsoidal shell (Figure 11.2) is a perfectly reflecting mirror surface. Two spheres of different radii, at equal temperature T, are placed at the two foci. Each sphere radiates light, as any body above absolute zero does. Of the two spheres at the same temperature, the larger one radiates more energy (assuming they are of the same material and color). Since every ray emitted from one focus passes through the other after a reflection, all rays from the larger sphere hit the small one, and vice versa. It follows that the larger sphere gives more than it gets when the two temperatures are equal, and thus the small sphere will heat up while the large one will cool down. Now a temperature

---

[2] we assume the ellipse reflects light perfectly.

difference can drive an engine,[3] so that our device gives a perpetual supply of free energy.

Question. Where is the mistake?

Answer. Not every ray leaving the larger sphere will hit the smaller one. Some will return to the larger body; this applies, for instance, to the rays leaving the large body to the left in Figure 11.2. Furthermore, the rays leave the surface of the body in all directions, not just the radial ones, and those nonradial rays do not line up with the foci. This destroys the original argument.

---

[3] For example, by causing a convection and using the convective flow to spin a wheel.

✳12✳

SAILING AND GLIDING

Question. Is it possible to sail on a river on a windless day?

Answer. Yes, sailing with no wind is possible, thanks to the current; Figure 12.1 explains how. The sail in still air acts like a knife in butter: it can only slice through the air, moving along the line of the sail.[1] On the other hand, the keel is being pushed by the current, and thus the boat slides as shown, towards the shore at the right angle. So the keel now acts as the sail and the flowing water acts as the blowing wind. And the sail, slicing through the air, acts as if it were a keel! Its just like regular sailing, except upside-down.

The sail–keel symmetry. We just discussed the boat from the point of view of a shore observer. But imagine yourself on that boat; you will then think that the water is still and that the wind is blowing upstream instead. So you would be in a conventional situation of a boat sailing in the wind. To you the sail will act as a normal sail catching the wind, and the keel will be a normal keel slicing though water. This

---

[1] This is an approximation—of course the sail can move in the direction perpendicular to the line S, but we ignore this smaller motion.

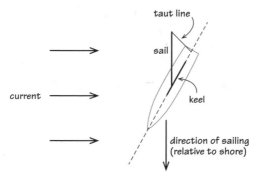

Figure 12.1. The air is still. The sail and the air swap the roles with the keel and the water.

is a neat symmetry: the sail and the keel exchange roles, depending on your reference frame! So the sail and the keel have completely equal rights in that respect.[2]

Problem. What are the possible directions in which the boat on the river can sail? The air is still and the river flows.

Solution. In a conventional situation (still water, blowing wind) the boat can sail in any direction not too close to the wind. Now our boat on the river is in exactly the same situation, except that the keel is acting as the sail, and vice versa. So the sail (and hence the boat) can move in any direction not too close to the direction upstream.

12.1 Shooting Cherry Pits and Sailing

One of childhood's pleasures is to eat a cherry and then have a cherry pit fight by squeezing the pit between two fingers and letting it shoot out with great speed.

---

[2] There is, however, one difference: the keel is aligned with the hull; the sail isn't. It is convenient perhaps to think of keel+hull as one, so that the boat consists of two units: (1) keel/hull and (2) sail.

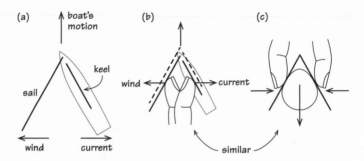

Figure 12.2. (a,b) The wind and the current try to pry open the sail–keel wedge, causing the wedge to slide in the direction of the arrow. (c) When the wedge is squeezed, it slides in the opposite direction.

Question. What is the similarity between the force propelling the cherry pit and the force propelling the sailboat?

Answer. Figure 12.2a shows the wind blowing against the river's current; the speeds of the two are equal. A sailboat is moving perpendicular to the shore. Now the wind and the current are like the two fingers pushing the sail and the keel, as Figure 12.2b shows. The "fingers" move apart, causing the wedge to slide as shown. Essentially the same thing happens when we shoot a cherry pit, except that the fingers squeeze, rather than push apart.

The importance of the reference frame. By assuming that the air and the water move with equal and opposite speeds we essentially chose the frame moving with the average speed of the wind and current. When we assumed that the air is still, we in effect attached our reference frame to the air. All this is summarized in Figure 12.3. Each time we chose a new frame of reference, we gained a new insight. The same happens in life.

134

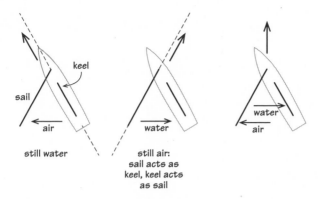

Figure 12.3. Sailing viewed from three different reference frames

12.2 Sailing Straight into the Wind

Question. A sailboat can sail in any direction that's not too close to the wind. A boat can get from point *A* to point *B* straight upwind by zigzagging (referred to as *tacking*). It would, however, be nice to be able to go *straight* into the wind. How to design such a boat?

Answer. Figure 12.4 shows a boat that will go against the wind. This boat will even automatically do so, adjusting its direction if the wind changes.[3] The design consists simply of a large wind propeller attached by a shaft to a smaller underwater propeller. With the right choice of propeller sizes, the wind will spin the air propeller, which will in turn make the water propeller "corkscrew" the boat straight into the wind. And because the big propeller is mounted on the stern, the boat will always want to head into the wind.

---

[3]   See   http://blog.modernmechanix.com/2007/11/27/wind-propeller-sails-proposed-for-liners/ for a more sophisticated design.

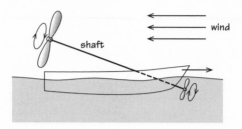

Figure 12.4. This boat will automatically point into the wind, and will travel against the wind.

12.3 Biking against the Wind

Biking into a stiff wind can be tough. Could a forward-facing windmill attached to the bike help? The energy generated by the mill could be used to assist in pedaling. The idea is appealing in some ways (and perhaps appalling in others). Casting aside practical issues, would it work in principle?

Let's consider an arrangement of Figure 12.5: the wind spins the propeller connected to an electric power generator that drives the electric motor which helps me pedal. *The generator, the motor, and the windmill are assumed perfect.* In particular, if I drag the windmill through still air, then the

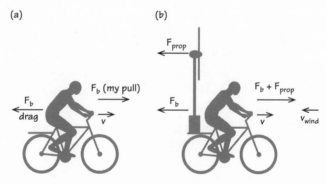

Figure 12.5. A windmill on a bike.

136

entire wattage I spend on dragging comes back to me in the form of electric energy. In other words, in still air such a perfect windmill on a bike would be a wash.[4]

But the wind carries energy, can it be used?

Question. Do I get an energy advantage from the windmill when biking against the wind?

Answer. To see if the windmill helps, let us calculate the difference between the powers I spend with and without the windmill,[5] while biking against the wind at the same fixed speed v.

1. *Without the windmill,* if I drag the bike through the air with constant speed v, I must apply the forward force equal and opposite to the drag force F_b felt by the bike + rider.[6] The power I spend is thus

$$F_b \, v. \tag{12.1}$$

2. *With the windmill,* I have to drag the propeller in addition to the bike + rider, thus applying greater force than before: $F_b + F_{\text{prop}}$, where F_{prop} is the force the propeller feels from the wind. The power I am expending now is

$$(F_b + F_{\text{prop}})v, \tag{12.2}$$

which is greater than before. But now I am getting the benefit of the energy generated by the windmill. How

---

[4] In practice, of course, it would be a losing proposition; instead of energy it may generate only laughs.

[5] Mechanical power is, by definition, the work done per second. Now the work is (by the definition) the force I am applying times the distance I moved in the direction of this force. Thus the power is given by the force times the speed in the direction of application of this force.

[6] This is the combined drag from the air and the tires.

much? Since the windmill is traveling through the air with speed $v + v_{wind}$, the power it generates is $F_{prop}(v + v_{wind})$, according to our earlier assumption of no losses. So the actual power required of me is the difference between what I spend and what I get:

$$(F_b + F_{prop})v - F_{prop}(v + v_{wind}) = F_b v - F_{prop} v_{wind}.$$

Comparing this with (12.1), the windmill lets me spend less power by the amount

$$F_{prop} v_{wind}.$$

We discovered:

The windmill mounted on a bike traveling into the wind[7] benefits the biker by the same amount of power as produced by a stationary windmill.

Question. Assuming that a spinning windmill propeller offers the same air resistance as a billboard, how are the wakes behind the two obstacles different in terms of their kinetic energies?

Answer. The windmill converts some of the kinetic energy of the air into electric energy. The billboard leaves much more of the kinetic energy of the air intact. Thus the air behind the windmill is calmer (in the sense of having less kinetic energy) than in the wake of the billboard.

12.4 Soaring without Updrafts

Question. Can a glider keep climbing steadily with no updrafts? By "no updrafts" I mean that the air is moving only in the purely horizontal direction.

---

[7] Subject to all of the simplifying assumptions mentioned before.

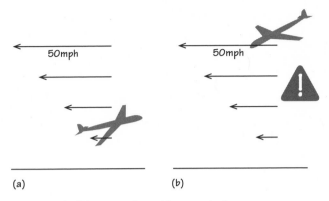

Figure 12.6. A glider ascending without updrafts.

Answer. Yes, at least in principle—for example, if the wind speed changes with altitude.

How does it work? Let's launch a glider into the shear wind, with the nose pointing slightly up, Figure 12.6. In still air the climbing glider would lose its airspeed and drop. But in the shear wind, a remarkable thing happens: as the glider rises, it enters the region of faster headwind, thanks to the shear. By climbing, the glider simply inserts itself into the ever faster moving air, thereby maintaining its airspeed—with no engine. Of course, the glider also is blown "backwards" in the process. But once it gains speed in can descend and return to the point of origin, circulating indefinitely. Some sea birds use this mechanism to soar without updrafts.

Smart birds. As an interesting aside, ocean waves cause the air to move up and down, creating very temporary updrafts—as the crest of the wave approaches, the air rises with water surface—and downdrafts—as the crest of the wave departs, the air descends. These are not the steady thermal updrafts

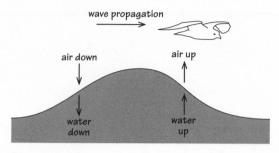

Figure 12.7. Riding a wave amounts to staying above the moving updraft zone.

that occur over land. But if a bird follows the wave, staying ahead of the crest, it stays in the updraft area, thus gliding indefinitely without flapping its wings. Pelicans, albatrosses, and some other sea birds do exactly this. They do just what human surfers do, except that they do it without touching the water! Birds' wings act as airborne surfboards. This also gives another way to think of surfers: they stay on that part of the wave which moves up.

Can a hang glider airsurf the waves? It is conceivable that a hang glider could air surf some enormous waves—although not very likely. One could, in principle, glide indefinitely, as Figure 12.8 illustrates, riding along one crest, rising before the wave breaks, joining the next incoming wave, and so on. I would try this with a remote-controlled glider first.

Question. A surfer is riding a wave with a constant speed. What does he have to do to speed up? Assume that the wave travels with constant speed without changing its shape.

Answer. To speed up, he should decrease the angle between his direction of motion and the line of the crest. This will put him on a steeper part of the wave and he will gain speed.

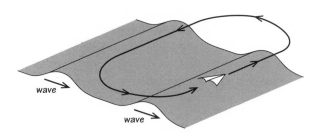

Figure 12.8. Airsurfing the waves.

12.5 Danger of the Horizontal Shear Wind

Gliding without updrafts sounds like an appealing idea, but it has a dangerous flip side. Imagine our glider in a shear wind, as in Figure 12.6b. Let's say the glider has a low airspeed, close to a stall. Wishing to speed up, he naturally noses down; in still air it would work since he would gain airspeed. But in our case he may actually lose airspeed if the shear is strong enough, and he can stall.

Question. What will a parachutist feel in a wind shear?

Answer. When a parachute descends in still air, the parachutist doesn't feel the wind in her face. However, in a (horizontal) wind shear, the parachutist will feel as if a steady horizontal wind were blowing, as if she had an invisible propeller pushing her in a horizontal direction.

✸13✸

THE FLIPPING CAT AND THE SPINNING EARTH

13.1 How Do Cats Flip to Land on Their Feet?

A cat released with his feet pointing up needs only a fraction of a second to point his feet down. How does he do it, with nothing to push off of?[1] Some people say that the cat does it by spinning the tail. On closer inspection this turns out to be false. As an experimental fact, tailless cats are just as good as the tailed ones in flipping over. Alternatively, a theoretical argument shows that to accomplish a 180° flip in a fraction of a second, the cat would have to spin its tail so fast that its tip would have to break the sound barrier, or to come close. This would create a sonic boom, or a loud whistling at the very least. And the enormous centrifugal force would cause a part of the tail to tear off and become a deadly projectile, almost like a bullet. So the "tail" theory quickly flunks the sanity test.

What makes the cat's achievement surprising is that his spin, having started at zero, must remain zero, given that no

---

[1] It is not hard to do a similar experiment on oneself, by sitting in a swivel chair and trying to turn around without touching the ground.

bend   and start   keep   unbend
              twisting   twisting

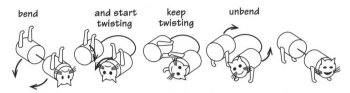

Figure 13.1. The main stages: (1) bending; (2, 3) twisting; (4) unbending. The cat's brilliant idea of getting by with zero angular momentum: when twisting, the spins of the two halves cancel, and so no net rotation is involved.

torques act on the cat while he is airborne.[2] So how does the cat manage to fool the zero-spin condition and flip?

Figure 13.1 shows the bare principle on a parody of the cat: two cylinders connected by a thin flexible waist. Starting with the feet pointing up, he bends his body in half as shown. Then he squirms his waist to counterrotate the two cylinders around their axes, until the feet point down. The cylinders rotate in the opposite directions, so that the net spin during the rotation is zero, as is required by the law mentioned before! Herein lies the cat's genius. Finally, the cat straightens his waist, and is done.

Incidentally, the "squirming" of the waist (steps 2 and 3 in the figure) does not twist it.[3] Had the cat not bent his body but kept it straight, this rotation would not have been possible; to turn his "thorax" around its axis, the straight cat would have have to counterturn his "pelvis" in order to

---

[2] The technical name for the spin is the *angular momentum*. In the absence of external torques on the flying cat, the cat's angular momentum is conserved, i.e., remains zero in flight. Some background on angular momentum can be found on page 175.

[3] Indeed, imagine holding a rubber hose shaped as a "U" with both hands. You can simultaneously twist the right end of the hose clockwise and the left end of the hose counterclockwise without twisting up the hose.

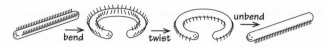

Figure 13.2. How a centipede could land on its feet. More feet, but the same principle.

maintain zero spin; this would cause the waist to twist like a corkscrew. Such cat would land on its feet, but it would be a twisted cat.

The idea is made even clearer by considering a centipede, as in Figure 13.2.

Although our parody of the cat is a bit extreme, it catches the gist of what real cats do: they bend their bodies when flipping—not by 180° of course, but perhaps by about 45°; and then they squirm their waists, just like our idealized cat did. But because they bend at a lesser angle, they have to squirm more times to achieve the desired flip.

13.2 Can Trade Winds Slow Earth's Rotation?

Question. Trade winds, which blow east, exert friction on the surface of the ocean. This friction has acted against Earth's rotation, over many millions of years. Could this friction have slowed down Earth's spin?

Answer. No, since the combined angular momentum of Earth and the atmosphere is conserved. This means that the total effect of atmospheric motion on Earth's rotation is nil. There are other winds (e.g., westerlies) acting in the opposite direction. Actually, rather than spinning up, Earth is slowing down its rotation, due mostly to the tidal braking of the Moon. It is estimated that Earth's day started out at 6.5 hours, having slowed down to the current 24 hours over the course of 4 or so billion years. The change is, of course, negligible over the course of human history.

As Earth passes its angular momentum to the Moon,[4] the Moon is moving away from Earth, a bit similar to the long satellite, as described on page 9.

---

[4] I am simplifying all this, ignoring the effect of the Sun and the much smaller effect of other planets.

⊠ 14 ⊠

MISCELLANEOUS

14.1 How to Open a Wine Bottle with a Book

This method actually works—I tried it myself, having been stimulated by a combination of scientific curiosity and the lack of a corkscrew, not necessarily in that order.

Begin by pressing a book against a wall. Then strike the bottom of the bottle against the book. I recommend holding the bottle with a towel and wearing protective glasses to prevent injury should the bottle break (anyone doing this with a champagne bottle risks winning Darwin's award). With repeated strikes, the cork will inch out, bit by bit, so that eventually it can be pulled out by hand.

I did this as a student a long time ago, back in the Soviet Union, sacrilegiously using a volume of Lenin's collected works. It was a good book—for my purpose, at least—although a corkscrew would have been better.[1]

Question. What drives the cork out of the bottle?

---

[1] The economic policies advocated in Lenin's volume led to shortage of many things, probably corkscrews among them, and caused that volume to be used instead of the corkscrew. This is a rare example of historical justice.

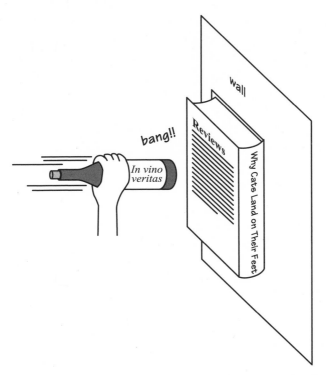

Figure 14.1. With each strike against the book the cork inches out. Why?

Answer. The short answer is "wine hammer"—an effect similar to the "water hammer" in plumbing, caused by a sudden stoppage of water in a pipe and causing occasional damage. Water hammer is also known as "hydraulic shock."

Figure 14.2 shows the "movie" of the process.[2]

In Figure 14.2a the bottle accelerates into the wall; because of this acceleration, the wine collects on the left,

---

[2] For full disclosure, this is my own opinion, and has not been verified by any measurements or direct observations, say by a high-speed camera. Few funding agencies would be excited to advertise their funding of such research.

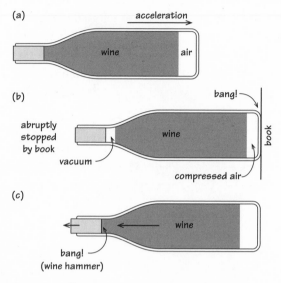

Figure 14.2. How the cork is getting hammered by wine out of the bottle.

against the cork (just like the passengers on a rapidly accelerating bus are thrown backward); the air is now between the wine and the bottom. Then, suddenly, the bottle is stopped by the book—but the wine keeps going by inertia, opening a bubble of vacuum near the cork (Figure 14.2b) and compressing the air on the right. The compressed air, acting as a spring, slows the wine down, and then *drives it back into the cork*. The vacuum bubble collapses—the vacuum, unlike the air, offers no cushion! At the moment of collapse, the wine hits the cork like a hammer hitting an anvil, without the softening cushion. The cork inches out a bit. In effect, we are hitting the cork from the inside, using wine as the hammer! After many strikes the cork protrudes enough to be pulled out by hand.

Cavitation. Collapsing bubbles of vacuum create great forces in other less desirable situations, for instance, next to boat propellers. If a propeller spins too fast, it can create a vacuum bubble. If this bubble then collapses next to the propeller, the resulting shock can pit the metal and damage the propeller.

These phenomena—the cavitation, the bottle opening, the hydraulic shock—are all manifestations of Newton's second law $F = ma$, where a huge a results in a huge F. Since electric currents have inertia[3] as well, a similar effect occurs in electric circuits. With this effect one can get a big electric shock from a small battery (see, e.g., Levi, *The Mathematical Mechanic*, pages 178–79.).

14.2 "It's *Alive!*"

Problem. If a weight is suspended from the ceiling on a collection of ropes and springs, and one of the springs is cut, will the weight drop lower? Specifically, what will happen to the weight in Figure 14.3 if the middle spring is cut?[4]

Solution. The weight will move up. Here is why. Let's hold the weight fixed and then cut S. Tensions of both ropes will increase as the result, tensions of springs will stay the same.

---

[3] Which in electricity is called induction.

[4] This paradox was originally described by Dietrich Braess in the context of traffic networks. Braess discovered that adding an extra road can actually increase travel time for all, (see "Ueber ein Paradoxen der Verkehrsplannung" (A paradox of traffic assignment problems), *Unternehmensforschung* 12 (1968), pp. 258–268. A mechanical experiment with springs similar to the one in Figure 14.3 – an analog of traffic networks – is described and studied in C. M. Penchina and L. J. Penchina, "The Braess paradox in mechanical, traffic, and other networks," *American Journal of Physics*, May 2003, pp. 479–482. I am grateful to Paul Nahin for pointing out these references.

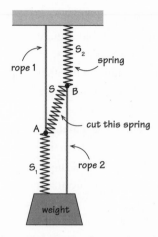

Figure 14.3. Will the weight move down when the middle spring is cut?

Hence the weight is now pulled up with greater force. So if we release the weight it will ascend as claimed.

Putting it differently, the spring S pulls the weight down by pulling point B down. Cutting this spring therefore moves the weight up.

14.3 Falling Faster Than g: A Falling Chain "Sucked in" by the Floor

The surprising "suction." This remarkable observation is due to Andy Ruina; a movie and a paper can be seen at http://ruina.tam.cornell.edu/research/topics/fallingchains/.

Holding the "rope ladder" sketched in Figure 14.4 by one end, you let it drop. An amazing thing happens: from the moment the chain comes in contact with the floor, the rest of the chain falls faster than a free-falling body; the ground seems to suck in the falling part of the chain. Why does this happen?

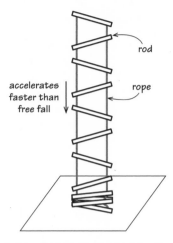

Figure 14.4. The falling chain is "sucked in" by the impacts with the floor.

An explanation. Let's understand what happens as an individual link hits the floor. The process is similar to the more familiar one of a pencil dropped on the floor. As one end of the pencil hits the floor, the other end jerks downward (assuming the pencil is not too vertical at impact). The chain links may experience a similar downward jerk if hitting the table at an appropriate slant, and will then pull the rest of the chain downward.

14.4 A Man in a Boat with Drag

The standard "man in a boat" problem goes like this. A man, standing at the stern of a resting boat, walks over to the bow. How far will the boat travel relative to the shore? The length of the boat and the man/boat mass ratio are given. The resistance of the water is to be neglected.[5] Now here

---

[5] We recap the solution. Let Δp and Δb be the displacements of the person and of the boat relative to the ground. We are interested in Δb. Since the center

is a remarkable twist on this question, told to me by Dima Burago.

Problem. Assume that the water exerts drag force F on the boat, directly proportional to the boat's speed: $F = kv$, where k is a (nonzero) constant. How far from its initial position will the boat end up after the man walks from one end of the boat to the other? Everything starts at rest. The two masses and the length of the boat are given (m, M, L).

Solution. The boat ends up where it started! The mass of the man does not matter, nor do the length and the mass of the boat, nor does the magnitude of the drag coefficient k. The problem requires *no* data!

A description of the motion. (A precise solution is given in the next paragraph.) When the person starts walking right, the boat starts moving left (Figure 14.5). Hence drag force points right, and by Newton's second law the center of mass

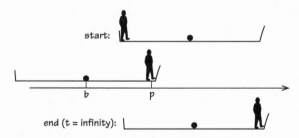

Figure 14.5. The boat eventually approaches its starting position.

---

of mass does not move, $m\Delta p = M\Delta b$. In addition, $\Delta p + \Delta b = L$. Solving the two equations for Δb we obtain $\Delta b = (m/(m + M))L$. This formula agrees with intuition: the heavier the boat, the less is its displacement; on the other hand, a very heavy person would make the boat move by almost its (the boat's) length.

of the man+boat accelerates right.[6] Having acquired motion to the right, the center of mass of man + boat continues by inertia even after the man sits down. Eventually, all slows down due to drag. To recap: the boat started moving left, activating the drag, which caused the center of mass of the whole system to move right—the motion that persisted once the man sat down. Remarkably, the boat will approach its initial position as time goes on. Why this strange coincidence happens is still unclear from the argument just given, but it is explained in the next paragraph.

The justification of the remarkable answer is quite simple, but it requires a little calculus. Let $b = b(t)$ denote the position at time t of the boat's center of mass (all is measured in a reference frame of the shore), and, similarly, let $p = p(t)$ be the position of the person (treated as a point mass). The center of mass[7] of the boat–person system is the weighted average of the two positions: $C = C(t) = (mp + Mb)/(m + M)$. Newton's second law (stated on page 172), applied in the direction of the boat's motion, gives

$$(m + M)\ddot{C} = -k\dot{b};$$

here each dot denotes the time derivative. Substituting the expression for C we obtain

$$m\ddot{p} + M\ddot{b} = -k\dot{b}. \qquad (14.1)$$

Let us integrate this relation from $t = 0$ to $t = \infty$. The Fundamental Theorem of Calculus[8] gives $\int_0^\infty \ddot{p}\, dt = \dot{p}(\infty) - \dot{p}(0)$. But $\dot{p}(0) = 0$ because all starts at rest, and

---

[6] This is a little like a cartoon dog running almost in place: his feet slide backward on the ground (like the boat sliding in water), while the dog's center of mass accelerates forward. Cartoon characters, however, routinely break Newton's laws.

[7] See page 169 for the definition.

[8] See page 184 for the details.

$\dot{p}(\infty) = \lim_{t \to \infty} \dot{p}(t) = 0$ because all ends up at rest. We conclude that $\int_0^\infty \ddot{p} \; dt = 0$. Similarly, $\int_0^\infty \ddot{b} \; dt = 0$. Integration of (14.1) thus gives

$$0 = k(b(\infty) - b(0)).$$

This shows that the boat's final displacement $b(\infty) - b(0) = 0$ *provided* the drag coefficient $k \neq 0$. So $b(\infty) = b(0)$: the boat approaches it's initial position as $t \to \infty$.

Remarkably, this fact does not depend on the size of k, as long as $k \neq 0$. However, what the size of k does affect is the rate of approach to that position. The smaller k, the longer it takes to come close to the final position. And for $k = 0$ this approach never happens.

14.5 A "Phantom" Boat: No Wake and No Drag

Before the main puzzle, here is a warm-up question.

Question. Would a boat experience drag if water was perfectly nonviscous?

Answer. It is the waves that carry off most of the energy generated by the engine, not viscosity. The smaller the waves the boat leaves, the more efficient it is. A hull shaped so as to leave almost no waves would be extremely efficient.

Question. Is it possible to design a boat that would leave almost no wake, at least in principle? Neglect viscosity and assume a constant travel speed and no waves in the water to begin with.

Solution 1. (due to Andy Ruina). The hull is made of a honeycomb of pipes; the intake and the outlet of each pipe are aligned, as shown in Figure 14.6. The water enters a typical pipe at A and exits at B. Absent viscosity, such a

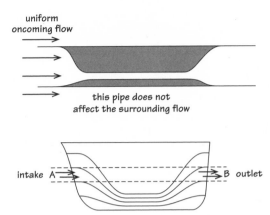

Figure 14.6. (a) The pipe is not seen by the surrounding flow. (b) A hullmade of many such pipes will leave no wake (in the ideal world).

hull is "invisible" to the water: if moving steadily through water, it would leave no disturbance behind.

Solution 2. To prevent the waves from forming, one could surround the conventional boat with a skirt—a disk flush with the water's surface, as in Figure 14.7. If the skirt is wide enough, almost no waves will be created. (While this may work in theory, it does not seem like a very practical solution for several reasons, not to mention the negative esthetic appeal of a boat with a tutu.)

D'Alembert's paradox and the boat with a skirt. In about 1752 D'Alembert realized that the drag is zero on a body moving with constant velocity relative to a fluid, provided that the fluid is ideal,[9] in the assumption that the fluid fills

---

[9] That is, incompressible, inviscid, irrotational. Here is a brief explanation of these terms. "Inviscid" means "lacking viscosity"; "irrotational" means zero

(a)   (b)

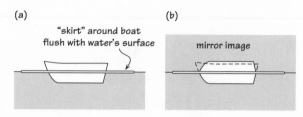

Figure 14.7. The skirt around the boat suppresses creation of the waves, thus decreasing drag.

the entire space. But let us replace the world above the surface in Figure 14.7 with the mirror image of the world below the surface; the water now fills the entire space and the symmetric "boat" becomes a submarine; according to D'Alembert's paradox, it will feel no drag (subject to the idealized conditions mentioned in the footnote). One can argue that the effect of the skirt is to make D'Alembert's paradox applicable, or nearly so.

14.6 A Constant-*G* Roller Coaster

Can a roller coaster track be shaped to make a rider feel a constant G-force, say, $2g$, where g is the grawitational acceleration?

Problem. The track is a curve in the vertical plane; the cart is a bead sliding on this curve with no friction, subject to gravity.

Solution. Yes. For any G-force $G > g$ there exists such a track, as in Figure 14.8. With a proper initial speed, the rider will feel as if he weighs mG if his weight on the

---

vorticity (page 52). The details can be found in any book on fluid dynamics, e.g., in Batchelor's classical text, *An Introduction to Fluid Dynamics*.

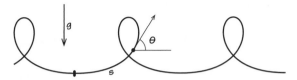

Figure 14.8. Coasting along a track described by Kepler's equation, we would experience a constant G-force, assuming we launch with an appropriate speed.

ground is mg. Note the high curvature nearer the top—this is necessary to increase the centrifugal force, to compensate for the double whammy of (1) less centrifugal force due to slower speed when higher up and (2) gravity trying to pull the rider out of the seat.

Kepler's Equation. Interestingly, the angle θ with the horizontal for such a constant g-force track satisfies Kepler's equation

$$\theta - e \sin\theta = cs, \qquad e = g/G < 1,$$

where c is a constant related to the speed at the lowest point and s is the arclength measured along the track. By choosing a different constant c we get different sized roller coasters, all with the same G. Showing how Kepler's equation comes up in the roller-coaster problem requires some calculus and is omitted here.

Interestingly, Kepler's equation arose originally in astronomy, in a completely unrelated setting.

14.7 Shooting at a Cart

Question. A drum is mounted on a cart, as shown in Figure 14.9. The drum can spin, and the cart can roll, without

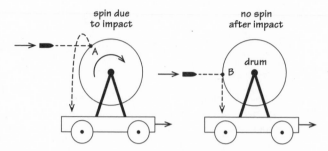

Figure 14.9. How much faster will the second cart roll?

friction. Let us do two separate experiments. In the first, the bullet hits the drum at point *A*, causing the drum to start spinning, and falls into the tray. The whole thing, bullet resting in the tray, starts rolling. The second experiment is identical except that the bullet hits at *B*, so that now the drum does not spin. With no energy spent on the spin, more energy is left for other motion. How much faster will the second cart move? Take the masses of the bullet and of the drum to be equal, and neglect the mass of the carts. You can assume that all of the drum's mass in in its rim.

Answer. Two carts will move with the same speed! I hid the mistake by saying, correctly, that "more energy is left for other motion" in case 2. But what I didn't say was that this "order motion" is not in the faster rolling, but rather in the extra heat produced at impact of the bullet hitting the drum head-on. In short, the rotation in case 1 has exactly the same energy as the extra heat produced in case 2.

An explanation of the answer. Conservation of linear momentum explains why the rolling speeds are the same in both cases. The main point is that spinning contributes zero to the drum's linear momentum, since each particle's momentum is canceled by the equal and opposite momentum of its antipodal counterpart. So whether or not the drum spins after the impact, the linear momentum of the entire assembly (bullet + cart + drum) is just Mv, where M is the total mass and v is the final speed. But this momentum all came from the bullet's momentum:

$$Mv = mV,$$

where m is the bullet's mass and V is the bullet's velocity before impact.[10] This shows that v, the cart's speed, does not depend on the spin.

14.8 Computing $\sqrt{2}$ with a Shoe

Problem. Given a stopwatch and a sneaker, can you find an approximate value of $\sqrt{2}$?

Solution.

1. Hang the sneaker on its lace and you have a pendulum. Using a stopwatch, record the number of swings in one minute; call this number n_1.
2. Fold the lace *in half* and record the new number of swings in one minute; call this number n_2.
3. The answer: $\sqrt{2} \approx n_2/n_1$.

For better precision take longer than one minute (and use a smaller sneaker).

---

[10] I am using capitals (M, V) for the larger quantities and lowercase letters (m, v) for the smaller ones.

The explanation of the method is very simple. The time T of one full swing of the pendulum is given by

$$T = 2\pi \sqrt{\frac{L}{g}},$$

where L is the length of the string and g is the acceleration due to gravity.[11] I suggested using a small sneaker since the formula assumes a point mass for the pendulum. For the periods of two pendulums of lengths L_1, L_2 we get therefore

$$\frac{T_1}{T_2} = \sqrt{\frac{L_1}{L_2}}.$$

Now we took $L_1 = 2L_2$, and counted the number of swings per minute for each length. We have $T_1 \approx 1/n_1$ minutes, since there are n_1 whole swings in a minute. Similarly, $T_2 \approx 1/n_2$. Therefore

$$\frac{n_2}{n_1} \approx \frac{T_1}{T_2} = \sqrt{\frac{L_1}{L_2}} = \sqrt{2}.$$

This also suggests how to compute other square roots. For example, to compute $\sqrt{3}$ make the lace 3 times shorter, so that $L_1/L_2 = 3$.

---

[11] Strictly speaking, this formula is an approximation—but it is a very good one for small-amplitude swings.

APPENDIX

The appendix is a short primer on concepts referred to in the book.

A.1 Newton's Laws

Newton's laws are all stated in an *inertial* reference frame, that is, in the frame that moves with no acceleration and no rotation.

Newton's first law. A body continues moving with constant speed in a straight line or stays at rest for as long as the sum of all forces acting on the body is zero.

Newton's second law. Forces applied to a body cause acceleration, and this acceleration **a** is in direct proportion to the sum of applied forces **F**:

$$m\mathbf{a} = \mathbf{F}; \tag{A.1}$$

the coefficient of proportionality m is referred to as the *mass*. Here **a** and **F** are vectors, Figure A.1. Often we look at the projection of Newton's law (A.1) onto a particular direction; for example, when speaking of the rectilinear motion we only care about the direction of the line. In such cases we treat the acceleration and the forces as scalar (as opposed to vector) quantities.

Figure A.1. Newton's second law.

According to Newton's law, $m = F/a$ (for the scalar case); in particular, if the acceleration $a = 1$, then $m = F$; that is, the *mass is the force required to give the body a unit acceleration*. This is precisely how we get an intuitive feel of inertia: we see how much force is needed to accelerate the body.

Newton's first law is a special case of Newton's second law, and, in my opinion, it is stated separately only because it is such an important special case.

A common mistake in applying Newton's laws lies in omitting some forces from the resultant force $\mathbf{F}$ in (A.1); several of our paradoxes (2.1, 4.2, and 4.4, for example) are based on this mistake.

Newton's third law. Two interacting objects exert equal and opposite forces upon each other: if object A applies force $\mathbf{F}$ to object B, then B applies force $-\mathbf{F}$ to A (Figure A.2).

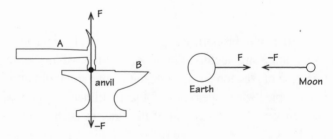

Figure A.2. Newton's third law.

Problem. While dragging a box along the floor, the force with which I pull the box forward is equal to the force with which the box pulls me backward. How is it then that I win, and not the box?

Solution. The implicit mistake in the last question is in the failure to pay attention to *all* the forces acting on me (and the same applies to the box). If I drag with constant speed, then the friction of my feet with the ground is in balance with the pull upon me from the box. And when I *start* dragging the box, that is, when I accelerate, the friction is greater than the box's pull, so I accelerate. And for the box, my pull on it is greater than its friction with the floor, so the box accelerates as well. At no time in this discussion did we have to compare my force upon the box with the box's force upon me.

A.2 Kinetic Energy, Potential Energy, Work

Since both kinetic and potential energies are defined by work, we define work first.

A.2.1 Work

Consider a constant force F pushing an object a distance D (Figure A.3). We define the work W done by this force as

$$W = FD. \tag{A.2}$$

What if the force is not lined up with the direction of motion? In that case, we modify the above definition, by using only the component of the force in the direction of the line:

$$W = F_1 D = F \cos \theta D. \tag{A.3}$$

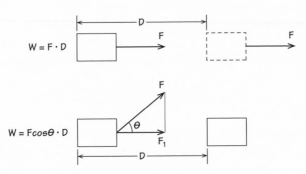

$$W = F \cdot D$$

$$W = Fcos\theta \cdot D$$

Figure A.3. Definition of mechanical work.

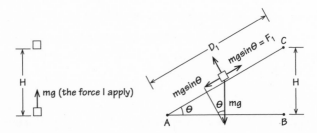

Figure A.4. Work required to move a mass from A to C in the gravitational field does not depend on the path.

Example. If I lift the weight $F = mg$ vertically up, to the height H, I must apply force mg over the distance H, thus doing work mgH, Figure A.4. If I drag the same weight up a ramp to the same height, I apply less force: $F_1 = mg \sin\theta$, but over the longer distance $D_1 = H/\sin\theta$; the work I do is then $W_1 = F_1 D_1$. Since $\sin\theta$ cancels, the work is still the same as when lifting: $W_1 = W = mgH$. In a way, this is unfortunate, since otherwise we would have had a perpetual motion machine. Indeed, imagine for a moment that $W_1 > W$. We could then place a hybrid car at C, roll

164

down to A, thereby charging the battery with the energy W_1 (assuming no losses), then drive from A to B (this takes zero work since the road is flat), then ride the car elevator from B to C, spending energy $W < W_1$. We would have completed the cycle and ended up with energy surplus $W_1 - W$—too good to be true.

How to define work in the general case, when the force is not constant, changing along the path, and when the path itself is not straight? In this case the work is defined simply by breaking the path into many short segments. A short segment is nearly straight, and the force along it is nearly constant, so that the definition (A.3) applies to each such segment with good precision. One then adds up all these bits of work. The finer the break-up of the path, the closer the total is to what we mean by true work.[1]

A.2.2 Kinetic Energy

Kinetic energy of a mass moving with speed v is defined as the *work required to bring the mass from rest to speed v.*[2]

This definition implies, as we now show, that $K = mv^2/2$. Let us accelerate a mass m from rest to speed v, applying some constant force F (the value of F will cancel out in the end). The kinetic energy is defined as the work

---

[1] Formally,

$$W = \int_C \mathbf{F} \cdot d\mathbf{r} = \int_C F \cos \theta \, ds, \tag{A.4}$$

where s is the arc length along the path C (C stands for "curve").

[2] This definition presumes silently that the work is independent of how it's done: by a weak force over a long time, or by a stronger force over a short time, or even by a variable force. In this case the presumption is correct: the work does not depend on how it's done. This can be seen by breaking the time into short intervals and adding up the work done over each interval. The sum "telescopes," and one ends up with the same answer that we get in the assumption of the constant force.

done:

$$K = F \cdot D, \tag{A.5}$$

where D is the distance traveled and F is the constant force we are applying. Notice that we need to express this answer in m and v alone, so that F and D will hopefully cancel out in the end. We have

$$F = ma = m\frac{v}{T},$$

where T is the time it takes to reach speed v, and $D = v_{\text{average}}T = ((0 + v)/2)/T = (v/2)T$. Substituting the last two expressions into (A.5) gives

$$K = F \cdot D = \left(m\frac{v}{T}\right) \cdot \left(\frac{v}{2}T\right) = \frac{mv^2}{2}.$$

We now see why v is squared: both the force $F = m(v/T)$, and the distance $D = (v/2)T$ in (A.5) are proportional to v: for a fixed T, greater v requires a greater force F *and* also is accompanied by a longer distance D traveled—a "double whammy" which explains the square. The denominator 2 in $mv^2/2$ comes from the fact that $v_{\text{average}} = v/2$.

A.2.3 Potential energy

Potential energy of an object located at a point A in a force field is defined as the work required to bring the object to point A from some reference point O. In other words, it is the work we have to do *against* the force field, going from O to A.

Example 1. Let us choose the reference point O to be a point on the ground level. The potential energy of the mass located at point A at height H is, according to the above definition, the work required to bring the mass from O to A. This work is mgH, as was explained on page 164.

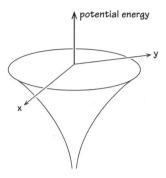

Figure A.5. Potential energy of a comet in a star's gravitational field.

Example 2 (requires calculus). For a comet in the Sun's gravitational field, let O be a point at infinity, and let the comet be at the distance r from the Sun's center. The potential energy of the comet is then the work we have to do against the gravitational force $F = k/x^2$ (here x is the distance to the Sun's center) as x varies from ∞ to r:

$$P(r) = \int_{\infty}^{r} \frac{k}{x^2}\, dx = -\frac{k}{r}. \tag{A.6}$$

The minus sign reflects the fact that when bringing the mass from ∞ to r we must apply the force *opposite* to the direction of motion. In other words, the gravitational field does the work for us when we are moving in from infinity. (To connect this to the previous example, the minus sign in (A.6) is similar to saying that the potential energy of a mass below floor level is negative.)

Figure A.5 shows the funnel-shaped graph of the potential energy of the comet.

The potential energy is defined up to an additive constant, due to the freedom of the choice of the reference point O. For instance, we can compute the potential energy of a ball

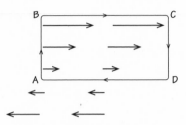

Figure A.6. An example of a nonconservative field.

in the room relative to the floor level, or, if we prefer, relative to the level of the tabletop, and so on; the results will differ by a constant.

Conservative force fields. Our definition of potential energy involved a presumption that the work is independent of the path chosen between O and A. This presumption holds for gravitational and electrostatic fields. Such fields are called *conservative*. But there are other force fields in which work does depend on the path; Figure A.6 gives a simple example.[3] For such nonconservative fields the concept of potential energy makes no sense. One can extract energy out of a nonconservative force field. Referring to Figure A.6, the work done by the field over the closed path $ABCDA$ is positive. If such a force field could have been created by some fixed arrangement of charges, we would have had an unlimited source of energy.

A.2.4 Conservation of Energy

Let us consider a particle of mass m moving under the influence of a *conservative* force field (such as a comet around the Sun, or a projectile on Earth, neglecting the air

---

[3] In fact, mathematically speaking, conservative fields are exceptional: they are the ones whose curl vanishes.

resistance). As the particle moves, both its kinetic energy K and its potential energy P change individually; however, their sum remains constant in time:

$$K + P = \text{const.} \qquad (A.7)$$

This is a consequence of Newton's second law, coupled with the assumption that the force field is conservative.[4]

For the comet, the conservation of energy gives

$$\frac{mv^2}{2} - \frac{k}{r} = E = \text{const.}$$

In particular, if r decreases, then v increases, in agreement with intuition.

A.3 Center of Mass

The concept of center of mass was known and used as early as 2,400 years ago by Archimedes, and possibly before. The center of mass can be defined as the balance point of a body

---

[4] Here is a quick calculus proof of (A.7) in the scalar case. Multiply both sides of Newton's second law, $ma = F$, by the velocity $v = \dot{x}$ (dots indicate time derivatives):

$$m\ddot{x}\dot{x} = F(x)\dot{x}.$$

But this is equivalent to

$$\frac{d}{dt}\left(m\frac{\dot{x}^2}{2} - \int_0^x F(s)\,ds \right) = 0.$$

Now $-\int_0^x F(s)\,ds = \int_0^x (-F(s))\,ds$ is precisely the potential energy, i.e., the work I must do against the force F to bring the mass from 0 to x; the minus sign is due to the word "against." We conclude that

$$\frac{mv^2}{2} + \int_0^x (-F(s))\,ds = K + P = \text{const},$$

as claimed.

suspended in a constant gravitational field.[5] However, the center of mass is a purely geometrical concept and can be defined, without reference to physics, as the average position of the body's particles. For a simple example, consider a dumbbell consisting of two masses m and M placed on the x-axis, with coordinates x and X, respectively. Let m and M be integer; we then can think of m pennies at location x and M pennies at location X. To find the average coordinate we just add up the coordinates of all the pennies and divide by the number of pennies:

$$
\text{C.M.} = \frac{\overbrace{x + \ldots + x}^{m} + \overbrace{X + \ldots + X}^{M}}{\underbrace{m + M}_{\text{number of pennies}}} = \frac{mx + MX}{m + M}.
$$

In the general case of N masses m_i, $1 \leq i \leq N$, each at the position $\mathbf{x}_i$ in space, the position vector of the center of mass is given, in the same way, by

$$
\bar{\mathbf{x}} = \frac{1}{m} \sum m_i \mathbf{x}_i, \quad m = \sum m_i. \tag{A.8}
$$

Another way to think of this expression comes from rewriting (A.8) as

$$
\bar{\mathbf{x}} = \sum \frac{m_i}{m} \mathbf{x}_i;
$$

the position of the center of mass is the weighted average of the positions of particles, with weights taken according to the proportion each mass takes up of the total mass.

---

[5] The balance point in a variable gravitational is not defined, since the balance point depends on the body's orientation in such a field.

A.4 Linear Momentum

One particle. A constant force **F** applied to a mass m produces a constant acceleration **a**, with

$$m\mathbf{a} = \mathbf{F}. \qquad (A.9)$$

In time Δt the mass will change its velocity by $\Delta \mathbf{v} = \mathbf{a}\Delta t$, since **a** is the change of velocity per unit of time.[6] Multiplying both sides of Newton's law (A.9) by Δt and using $\mathbf{a}\Delta t = \Delta \mathbf{v}$, we get

$$m\Delta \mathbf{v} = \mathbf{F}\Delta t, \quad \text{or} \quad m\mathbf{v}_2 - m\mathbf{v}_1 = \mathbf{F}\Delta t. \qquad (A.10)$$

The vector $m\mathbf{v}$ is called the *linear momentum* of a mass. Intuitively, $m\mathbf{v}$ tells us the amount and the direction of motion.

In most of our examples, the motion takes place along the line, and in such cases we treat the momentum as scalar.

Problem. A door is slightly ajar. A bullet fired through the door will not move it much, despite the enormous force a bullet applies to the wood while passing through it. But a light pressure from a finger will open the door. How to explain this?

Solution. The finger gives the door more momentum because of the much, much longer duration of contact:

$$F_{\text{finger}}\Delta t_{\text{finger}} > F_{\text{bullet}}\Delta t_{\text{bullet}}.$$

(In this example the direction of the momentum is not in question, so we treat the momentum as a scalar.) Subconciously, we use the same effect when tearing off paper from

---

[6] For nonconstant $\mathbf{a} = \mathbf{a}(t)$, the above formula has to be modified: $\Delta \mathbf{v} = \bar{\mathbf{a}}\Delta t$, where $\bar{\mathbf{a}}$ is the average acceleration defined as $\bar{\mathbf{a}} = (1/\Delta t)\int_{t_1}^{t_2} \mathbf{a}(t)\ dt$; here $\Delta t = t_2 - t_1$.

a toilet paper roll: a quick jerk (as opposed to a gentle pull) will tear off the paper barely causing the roll to spin. Toddlers who haven't yet learned this skill can unwind the entire roll while trying to tear off one square.

Many particles. So far we discussed Newton's laws for just one point mass. But any complex system, such as a dumbbell with two masses, or the space shuttle, or even a cat, can be thought of as a collection of many point masses. Each mass may interact with others, and also be subject to an external force. Now *the center of mass of any collection of particles behaves as a single point mass, in the sense that it obeys Newton's second law*[7]

$$\mathbf{F} = m\bar{\mathbf{a}}, \tag{A.11}$$

where m is the total mass, $\mathbf{F}$ is the sum of all external forces, and $\bar{\mathbf{a}}$ is the acceleration of the center of mass. Importantly, $\mathbf{F}$ does not include internal forces, that is, the forces of interaction between particles of the system; as it turns out, these forces cancel out due to Newton's third law.

Proof of (A.11). This amounts to adding up the statements of Newton's law for each constituent particle and then using Newton's third law to cancel the forces of interaction between particles. The ith particle is subject to an external force, as well as to the sum of forces from all other particles except for itself:

$$m_i \mathbf{a}_i = \mathbf{F}_i^{\text{ext}} + \sum_{j \neq i} \mathbf{F}_i^j;$$

---

[7] Which we postulated only for a point mass.

here $\mathbf{F}_i^j$ is the force upon the ith particle from the jth particle.[8] By Newton's third law, $\mathbf{F}_i^j = -\mathbf{F}_j^i$. Thus, when we add all of the above equations, all the mutual forces cancel since $\mathbf{F}_i^j$ and $\mathbf{F}_j^i$ each occurs once in the total sum. What's left is

$$\sum m_i \mathbf{a}_i = \sum \mathbf{F}_i^{\text{ext}} = \mathbf{F},$$

which is the same as our claim (A.11), by using (A.12) below.

Bunching up the particles. *The linear momentum of any system of particles equals the linear momentum of their center of mass, endowed with the combined mass of all the particles.*

Proof. The definition of the center of mass (A.8), slightly rewritten, says

$$m\bar{\mathbf{x}} = \sum m_i \mathbf{x}_i. \qquad (A.12)$$

This implies that

$$m\bar{\mathbf{v}} = \sum m_i \mathbf{v}_i;$$

and proves the claim: the left-hand side is the linear momentum of the center of mass endowed with the total mass; the right-hand side is the linear momentum of the system.

Newton's second law for many particles has an immediate corollary:

The law of conservation of linear momentum. *If the sum of external forces upon a system of particles is zero—* $\mathbf{F} = \mathbf{0}$*—then the combined linear momentum of the system remains constant.* In particular, then, the center of mass of the system either rests or moves with constant velocity.

---

[8] So the superscript indicates the source of the force.

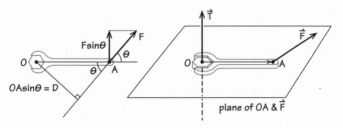

Figure A.7. Defining the torque.

A.5 The Torque

Consider a force F applied to a point A; another point O, called the pivot, is given.

The torque of F relative to the pivot O is, by definition, the product of the distance OA and the component of F perpendicular to OA:

$$T = OA \cdot F_\perp = OA \cdot F \sin\theta. \qquad (A.13)$$

The torque measures the intensity of turning. Now since $OA \sin\theta = D$ is the perpendicular distance from O to the line of the force, the torque can be rewritten as $T = F(OA \sin\theta)$, or

$$T = FD, \qquad (A.14)$$

where D is shown in Figure A.7. In other words, the torque can alternatively be defined as *the product of the force F and the distance D from the line of force to the pivot.* What we defined thus far is a scalar torque. Actually, the torque can be defined as a vector, as follows.

There is a natural "axis of rotation," the line perpendicular to the paper, that is, to the plane defined by the vector $\overline{OA}$ and by the force vector $\mathbf{F}$. Along this line we define a preferred direction—namely, the direction in which a nut with a right-handed thread would advance when turned by the force. The vector torque is the vector in that direction, of magnitude defined by (A.13). In other words, the torque is defined as the cross-product of the lever and the force:

$$\mathbf{T} = \overline{OA} \times \mathbf{F}. \qquad (A.15)$$

In fact, the above discussion motivates the definition of the cross-product.

A.6 Angular Momentum

Angular momentum M is a rotational counterpart of linear momentum. For a point mass m at P, the angular momentum relative to a point O is defined as $r(mv_\perp)$, where r is the distance to O and $v_\perp$ is the component of the velocity in the direction perpendicular to OP.

Actually, the angular momentum is endowed with a direction, and thus is a vector; what we defined so far is it's magnitude. The direction of this vector is taken to be the "axis of rotation,"[9] that is, the line perpendicular to both the position vector $\mathbf{r}$ and the velocity vector $\mathbf{v}$, as shown in Figure A.8.

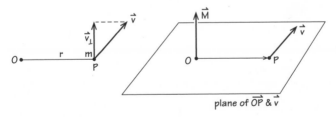

Figure A.8. Definition of the angular momentum.

---

[9] Although the mass need not be moving in a circle—it may move in a straight line or any other path.

Formally speaking,

$$\mathbf{M} = \mathbf{r} \times m\mathbf{v},$$

where $\mathbf{r}$ is the position vector, $\mathbf{v}$ is the velocity vector, and $\times$ is the cross-product. In fact, the two preceding paragraphs explain why the cross-product is defined as it is in texts on calculus and analytic geometry. It can be noted that the second factor in the above expression is the linear momentum. Thus *the angular momentum is the cross-product of the vectors of position and of linear momentum.*

For a system of many particles, the angular momentum is defined as the sum of the angular momenta of individual particles:

$$\mathbf{M} = \sum \mathbf{r}_i \times m_i \mathbf{v}_i. \tag{A.16}$$

Conservation of the angular momentum. *If the sum of external torques upon a system of particles is zero, then the angular momentum of the system is constant.*

In particular, if you don't touch the cat while it's airborne, the cat's angular momentum will remain unchanged, however the cat wriggles during the flight (neglecting the air resistance).

Before going to the proof, I would like to make the key point: the internal torques—the torques exerted by particles of the system upon each other—cancel, and so when adding up all the torques on all the particles, only the external torques remain in the sum. The cancellation of internal torques is seen from Figure A.9. Consider the plane defined by the origin O and the two masses m_i and m_j. The forces $\mathbf{F}_i^j = -\mathbf{F}_j^i$ lie in that plane. The directions of the torques exerted by these two forces around $\mathbf{O}$ are clearly opposite (both are perpendicular to the paper). To show that these torques cancel we must only show that their magnitudes are

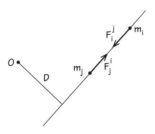

Figure A.9. Cancellation of internal torques.

equal. By (A.14), the magnitude of the torques are $T_{ji} = DF_{ji}$ and $T_{ij} = DF_{ij}$; it is important to note that the distance D (Figure A.9) is the same for both forces. Since $F_{ij} = F_{ji}$ by Newton's third law, the magnitudes or the torques are equal. This proves that the internal torques cancel.

We now proceed to the proof of the conservation of angular momentum. Wishing to show that $\mathbf{M}$ is a constant vector, we differentiate it by time in (A.16):

$$\frac{d\mathbf{M}}{dt} = \sum (\underbrace{\mathbf{v}_i \times m_i \mathbf{v}_i}_{0} + \mathbf{r}_i \times \underbrace{m_i \mathbf{a}_i}_{\mathbf{F}_i}) = \sum \mathbf{r}_i \times \mathbf{F}_i; \quad \text{(A.17)}$$

here $\mathbf{F}_i$ stands for the sum of all forces upon the ith particle, both from the external sources and from the other particles in the system, so that

$$\mathbf{r}_i \times \mathbf{F}_i = \mathbf{r}_i \times \mathbf{F}_i^{\text{ext}} + \sum_{j \neq i} \mathbf{r}_i \times \mathbf{F}_i^j.$$

The last term is the sum of all internal torques upon the ith particle. When we substitute this into (A.17), these internal torques add up and all cancel each other, according to the preceding paragraph. We are left with

$$\frac{d\mathbf{M}}{dt} = \sum \mathbf{r}_i \times \mathbf{F}_i^{\text{ext}} = \mathbf{T}, \quad \text{(A.18)}$$

where **T** is the sum of torques exerted by external forces. In the special case when $\mathbf{T} = \mathbf{0}$, we conclude that $\mathbf{M} = \text{const.}$ We proved that the angular momentum is conserved if the sum of external torques is zero.

Equation (A.18) is the rotational analog of Newton's second law.

A.7 Angular Velocity, Centripetal Acceleration

Angular velocity. For a point moving in a circle centered at point O, the angular velocity ω is defined as the *rate of change of the angle θ formed by the radius vector of the point and a fixed direction* (Figure A.10).

Angular velocity vs. speed. The speed of a point moving in a circle with angular velocity ω is given by

$$v = \omega r, \tag{A.19}$$

where r is the radius of the circle. This follows from the definition of the angular measure: recall that the radian measure θ of the angle of an arc is simply the length s of the arc divided by the radius of the circle, $\theta = s/r$, so that

$$s = \theta r.$$

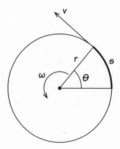

Figure A.10. Defining the angular velocity.

Now since s and θ are in direct proportion, then so are their rates of change, with the same coefficient; this proves (A.19).

Centripetal acceleration. Does a point moving with *constant* speed have zero acceleration? The answer is no, unless the point moves in a straight line. Acceleration can be caused by the change of direction of motion. Strictly speaking, the acceleration is the rate of change of the velocity *vector*, and thus is a vector itself. For a point moving in a circle with constant speed, the acceleration points toward the center and has the magnitude

$$a_c = \omega v. \tag{A.20}$$

This formula has the following explanation, beautiful in its brevity. The idea is to apply the formula $v = \omega r$—not to the physical circle, but rather to the circle traced by the tip of the velocity vector (parallel transported so its tail is at the origin at all times). Here are the details (Figure A.11).

With the tails of all velocity vectors **v** brought to the origin, the *tip Q of* **v** *moves in a circle of radius v with the angular velocity ω* (same as the angular velocity of the point, since **v** and **r** are perpendicular at all times). Thus, according

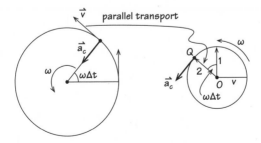

Figure A.11. Finding the centripetal acceleration.

to (A.19), applied to the velocity circle, we have

$$\underbrace{(\text{speed of } Q)}_{a_c} = \omega \underbrace{(\text{radius of the velocity circle})}_{v},$$

or $a_c = \omega v$, thus proving (A.20).

Equivalent formulas. Substituting (A.19) into (A.20) gives alternative expressions

$$a_c = \omega^2 r = \frac{v^2}{r}. \qquad (A.21)$$

Most texts use these expressions, but (A.20) is arguably more basic and slightly simpler.

Problem. As the car goes around the ramp, the tires grab the road with a certain force which keeps the car on the ramp. What happens to this force if you double the car's speed?

Solution. The force quadruples, according to (A.21).

Question. What is an intuitive explanation of this answer, without referring to a formula?

Answer. There is a double whammy: first, when you double the speed, you double the length of your velocity vector, and, on top of that, you double the rate of its turn. This quadruples the speed at which the tip of this vector moves; this speed is precisely the centripetal acceleration.

A.8 Centrifugal and Centripetal Forces

Imagine sitting in a carousel, traveling in a circle, or sitting in a car going around the highway ramp. It feels as if an invisible force were pulling you away from the center of the circle. This is a fictitious force, one could say, in the sense that no one is really pulling you away from the center—rather, it is an illusion caused by your tendency to go straight by inertia, in conflict with the veering of the car. This fictitious force is called the *centrifugal* force.[10] However, there is nothing fictitious about the force that *a passenger* applies to the car in the direction away from the center. This force could legitimately be called the centrifugal force— although this is not the force applied to the passenger's body.

The force applied to your body by the car causes you to travel in a circle. This force points at the center of the circle.[11] This real force is called the *centripetal* force. By Newton's second law, this force is given by ma_c, where a_c is the centripetal acceleration given by (A.21). We conclude that the centripetal force is given by

$$F_c = ma_c = \frac{mv^2}{r}.$$

A.9 Coriolis, Centrifugal, and Complex Exponentials

Background. Here I want to show how both the Coriolis and the centrifugal forces come out immediately by using complex numbers and simple calculus. All the required knowledge of calculus and of complex numbers is listed below.

1. *A complex number $a + ib$ is simply the point (a, b) in the plane; the points on the x-axis are identified with real numbers, so real numbers are a subset of complex ones; we thus write $(a, 0) = a$.

2. The angle between the positive x-axis and the position vector of (a, b) is called the *argument* of $a + ib$; the

---

[10] "Fugitive from the center": "effugere" is Latin for "to flee."
[11] Provided you travel with constant speed, i.e., neither speeding up nor slowing down.

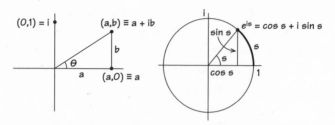

Figure A.12. Complex numbers, complex exponentials.

distance $\sqrt{a^2 + b^2}$ to the origin is called the *absolute value* of $a + ib$.

3. By the definition, two complex numbers are multiplied by adding their arguments and multiplying their absolute values. In particular, $i = (0, 1)$ has argument $\pi/2$ and absolute value 1; hence the argument of $i^2 = i \cdot i$ is $\pi/2 + \pi/2 = \pi$, and its absolute value is $1 \times 1 = 1$. Hence $i^2 = (-1, 0) \equiv -1$.

4. The complex exponential e^{is} is defined as the point $P(s)$ on the unit circle centered at the origin, at the distance s along the circle from the positive x-axis, Figure A.12. Recall that sin and cos are defined as the coordinates of $P(s)$. Thus, by the definition,

$$e^{is} = \cos s + i \sin s,$$

which is a famous formula due to Euler (he discovered many others).

How to rotate a point? If Z is a point in the plane, then $e^{i\theta} Z$ is the point obtained by rotating Z by θ around the origin. Indeed, the length of $e^{i\theta}$ is 1, and its argument is θ. Hence multiplying Z by $e^{i\theta}$ does not change the length of Z and adds θ to its argument, that is, turns Z by θ. We will need this remark shortly.

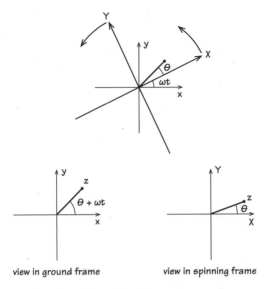

view in ground frame   view in spinning frame

Figure A.13. Coriolis and centrifugal forces via complex exponentials.

Coriolis and Centrifugal Forces in One Step. Let us imagine a person walking on a rotating platform. We have a coordinate system (x, y) drawn on the ground, and another coordinate system (X, Y) drawn on the platform, with the common origin at the center of the platform, Figure A.13. At $t = 0$ the two systems coincide; at $t > 0$ the platform's coordinate system turned by ωt. Let $Z = (X, Y) \equiv X + iY$ be a point on the platform, at time t. The same point in the ground frame will have turned by angle ωt around the center, so the point's position in the ground frame will be

$$z = e^{i\omega t} Z, \qquad (A.22)$$

according to the preceding paragraph. Let us compute the acceleration of the point. Complex exponentials are differentiated by the same rules as the real exponentials.

Differentiating twice, we get, after collecting terms,

$$\ddot{z} = e^{i\omega t}(\ddot{Z} + 2i\omega\dot{Z} - \omega^2 Z).$$

From this we get the *apparent acceleration* of the particle in the rotating frame:

$$\ddot{Z} = \underbrace{e^{-i\omega t}\ddot{z}}_{\text{actual}} \underbrace{-2i\omega\dot{Z}}_{\text{Coriolis}} + \underbrace{\omega^2 Z}_{\text{centrifugal}}.$$

So in addition to the actual acceleration, the rotating observer will feel the Coriolis and the centrifugal accelerations. Note that $i\dot{Z} \perp \dot{Z}$, so indeed the Coriolis acceleration is perpendicular to the velocity, confirming our intuitive discussion.

A.10 The Fundamental Theorem of Calculus

The theorem states that a function f with a continuous derivative, defined on an interval $[a, b]$, satisfies

$$\int_a^b f'(t)\, dt = f(b) - f(a). \qquad (A.23)$$

An essentially equivalent statement is

$$\frac{d}{dx}\int_a^x F(t)\, dt = F(x), \qquad (A.24)$$

for any continuous function F. Since the proof of the theorem is contained in every calculus book, I will omit it, giving instead an intuitive explanation of (A.23).[12]

Let us interpret $x = f(t)$ as the position of a point moving along the x-axis. Equation (A.23) simply gives two different ways to express the point's displacement during

---

[12] A geometrical explanation of (A.24) can be found, for instance, in Levi, *The Mathematical Mechanic*.

the time from time $t = a$ to time $t = b$. On the right-hand side of (A.23) is the displacement expressed as the difference between the ending and the starting coordinates of the point. On the left-hand side we have the sum of infinitesimal displacements. Indeed, the displacement during a short time Δt is given by $\Delta x = v\Delta t = f'(t)\Delta t$, since the velocity stays nearly constant during a short time.[13] The total displacement is then the integral, that is, the limit of the sums of these terms with $\Delta t \to 0$ (and with the number of terms growing to infinity).

---

[13] Since f' is assumed to be continuous, it barely changes over a short interval.

BIBLIOGRAPHY

V. I. Arnold, *Mathematical Methods of Classical Mechanics.* New York: Springer-Verlag, 1980.

G. K. Batchelor, *An introduction to Fluid Dynamics.* New York: Cambridge University Press, 1967.

D. Braess, *Ueber ein Paradoxon der Verkehsplannung* ("A paradox of traffic assignment problems), *Unternehmensforschung* 12 (1968). pp. 258–268.

L. C. Epstein, *Thinking Physics: Practical Lessons in Critical Thinking.* San Francisco: Insight Press, 1992.

H. G. Goldstein, *Classical Mechanics.* Reading, MA: Addison-Wesley, 1950.

A. Grewal, P. Johnson, and A. Ruina, A chain that accelerates, rather than slows, due to collisions: how compression can cause tension, *American Journal of Physics* 79, (7), July 2011; p. 723.

C. P. Jargodzki and F. Potter, *Mad about Physics: Braintwisters, Paradoxes, and Curiosities.* New York: John Wiley & Sons, 2001.

L. D. Landau and E. M. Lifshitz, *Mechanics*, 3rd ed. Oxford: Butterworth-Heinemann, 2002.

M. Levi, *The Mathematical Mechanic: Using Physical Reasoning to Solve Problems.* Princeton, NJ: Princeton University Press, 2009.

M. Levi, *Physica* D 132 (1999), p. 158.

P. V. Makovetsky, *Smotri v koren*, 3rd ed. Moscow, 1976.

M. Minnaert, *The Nature of Light and Color in the Open Air.* New York: Dover, 1954. Translated and revised edition, *Light and Color in the Outdoors*, New York: Springer-Verlag, 1993.

M. M. Michaelis and T. Woodward, *American Journal of Physics* 59(9) (1991) pp. 816–821.

J. Munkres, *Topology.* Upper Saddle River, NJ: Prentice-Hall, 2000.

P. Nahin, *Number Crunching. Taming Unruly Computational Problems from Mathematical Physics to Science Fiction.* Princeton, NJ: Princeton University Press, 2011.

Yu. I. Neimark and N.A. Fufaev, *Dynamics of nonholonomic systems.* Translated from the Russian. Providence, RI: American Mathematical Society, 1972.

C. M. Penchina and L. J. Penchina, The Braess paradox in mechanical, traffic, and other networks, *American Journal of Physics* (May 2003) pp. 479–482.

Ya. Perelman, *Physics for Entertainment,* Books 1 and 2. Moscow: Foreign Languages Publishing House, 1962–63.

E. J. Routh, *Dynamics of a System of Rigid Bodies,* Part 2, 4th ed. London: MacMillan and Co., 1884, pp. 299–300.

A. Stephenson, On a new type of dynamical stability, Manchester Memoirs 52 (1908), p. 110.

J. Walker, *The Flying Circus of Physics.* New York: John Wiley & Sons, 2007.

G. H. Wolf. *Physical Review Letters* 24 (1970). pp. 444–446.

INDEX